Mabel Loomis Todd

Corona and Coronet

Being a Narrative of the Amherst Eclipse Expedition to Japan...

Mabel Loomis Todd

Corona and Coronet
Being a Narrative of the Amherst Eclipse Expedition to Japan...

ISBN/EAN: 9783744746076

Printed in Europe, USA, Canada, Australia, Japan

Cover: Foto ©Andreas Hilbeck / pixelio.de

More available books at **www.hansebooks.com**

Corona and Coronet

BEING

A NARRATIVE OF THE AMHERST ECLIPSE EXPEDITION
TO JAPAN, IN MR. JAMES'S SCHOONER-YACHT
CORONET, TO OBSERVE THE SUN'S
TOTAL OBSCURATION
9TH AUGUST, 1896

BY

MABEL LOOMIS TODD

Author of " Total Eclipses of the Sun," etc., etc.

WITH ILLUSTRATIONS

BOSTON AND NEW YORK
HOUGHTON, MIFFLIN AND COMPANY
The Riverside Press, Cambridge
1898

TO

CAPTAIN AND MRS. ARTHUR CURTISS JAMES

THIS STORY OF ONE

CRUISE OF THE CORONET

IS AFFECTIONATELY INSCRIBED

PREFACE

THE expedition proper, sent out to observe the total eclipse of the sun in Japan on the 9th of August, 1896, through the liberality of Mr. D. Willis James and his son, owners of the schooner yacht Coronet, consisted of nine persons, — Captain and Mrs. Arthur Curtiss James; Professor and Mrs. David P. Todd; Passed Assistant Engineer John Pemberton, U. S. Navy; Mr. Willard P. Gerrish of Harvard College Observatory; Vanderpoel Adriance, M. D., of the College of Physicians and Surgeons, Columbia University; Mr. Arthur W. Francis of New York; and Mr. E. A. Thompson, mechanician, of Amherst.

Certain aspects of this memorable trip have seemed worthy of narration, covering, as it did, more than ten thousand miles of sailing for our party, and at least forty-five thousand miles of deep sea voyaging for the Coronet. As an "unscientific account of a scientific expedition," it necessarily makes divers small branchings in its course, like a sort of ornamental needlework much affected by our grandmothers. I have, as it were, feather-stitched my way to Yezo and back again.

To avoid repetition, our friendly company are named on paper as they were often designated on board, — the Captain, the Professor or Astronomer, the Doctor, the Musician, and so on. Mr. Francis was apt to be addressed as General, largely because of his masterful management of expedition finances; and Mr. Pemberton was known as Chief, having been many years chief engineer of the U. S. S. Monocacy while attached to our Asiatic squadron. During that time, he had in 1887 accompanied Professor Todd's earlier expedition to Shirakawa, in central Japan, where his assistance was peculiarly welcome.

The narrative owes much to many friends, first and foremost to the owners of the Coronet for making the expedition possible; to my husband for reducing to accuracy my attempts at describing the scientific phases of the trip; and to all our fellow voyagers for drawings, photographs, or material no less picturesque in its way. Of more than ordinary interest is the brief paper upon deep-sea sailing from the point of view of a practical and enthusiastic yachtsman, written by the younger of the Coronet's owners.

President Hill of the Great Northern Railway has put the whole expedition in his debt for the generous courtesy of transportation in his private car from Chicago to San Francisco. At the latter place Mr. Merrill and Mr. Wheeler were untiring in facilities accorded us, and their warehouses afforded most convenient headquarters for the expedition on the Pacific Coast.

In Honolulu, obligation was constant to the hospitable friends who united in showing us the characteristic side of Hawaiian life, as well as to others who gave practical aid to the more serious side of our work; especially to President Dole, and to Professor Alexander, surveyor general, and his assistants.

In Japan a list of those through whose kindness the pathway of the expedition was made smooth, even luxurious, would comprise almost every one with whom we came in contact, from personal friends of various nationalities who entertained us, to the government officials who granted railroad passes, special steamers, and facilities otherwise impossible. Many pleasant and essential favors were obtained through the friendliness of Mr. Hayashi and Mr. Kabayama; and of Mr. Herod, then chargé d'affaires of our legation at Tokyo. Also the governor of Hokkaido and the mayor of Esashi exerted themselves most courteously in our behalf.

Without the intelligent services of Mr. Oshima and Mr. Murakami, both teachers in government colleges, ease of communication in remote localities would not have been attainable; and to Professor Burton and Mr. Ogawa warm thanks are due for fine views of the Ainu and northern Yezo.

I must not omit mention of the kindly assistance in many technicalities, Hawaiian and Japanese, given me by Mrs. Frances Carter Crehore, formerly of Honolulu, and by Miss Umè Tsuda, of the Peeresses' School in Tokyo. Also the editors of "The Nation," "The Century Magazine," "The Atlantic Monthly," "The Independent," and "The Outlook," have kindly given permission to reprint my articles originally published in their magazines.

M. L. T.

OBSERVATORY HOUSE,
AMHERST, *October,* 1898.

CONTENTS

CHAPTER		PAGE
	INTRODUCTORY	xiii
	DEEP-SEA YACHTING BY A. C. JAMES	xxiii
I.	THE CORONET	1
II.	PREPARATION	8
III.	OVERLAND	14
IV.	SAUSALITO	24
V.	FIFTEEN DAYS AT SEA	30
VI.	LIFE IN HONOLULU	42
VII.	HAWAIIAN VOLCANOES	58
VIII.	A HAWAIIAN JOURNEY	68
IX.	KILAUEA	78
X.	A POI LUNCHEON	88
XI.	WITH KATE FIELD	97
XII.	A MID-PACIFIC COLLEGE	104
XIII.	THE LEPERS OF MOLOKAI	111
XIV.	FOUR WEEKS AT SEA	125
XV.	JAPAN REVISITED	139
XVI.	DEPARTURE OF THE EXPEDITION	155
XVII.	IN FAMILIAR HAUNTS	172
XVIII.	SOUTHWARD	181
XIX.	GIFU AND THE CORMORANT FISHING	188
XX.	KYOTO	194
XXI.	NARA	209
XXII.	YACHTING IN THE INLAND SEA	216
XXIII.	EXPEDITION EXPERIENCES	229

XXIV.	The Tidal Wave	241
XXV.	In Pursuit of a Shadow	254
XXVI.	Still Pursuing	264
XXVII.	Esashi in Kitami	272
XXVIII.	In Ainu Land	292
XXIX.	The Eclipse	318
XXX.	A Native Celebration	327
XXXI.	Voyage on a French Cruiser	336
XXXII.	Homeward Bound	343
XXXIII.	Back to an Arizona Copper Mine	357
	Index	377

LIST OF ILLUSTRATIONS

	PAGE
Mr. James's Schooner Yacht Coronet *Frontispiece*.	
Expedition Headquarters at Esashi. . *Facing*	10
The Rotary Snow-Plough at Wellington . .	20
Expedition Work on Board.	36
Residence of President Dole in Honolulu .	50
Hawaiian Village Landing-Place	64
Sulphur Blow-Hole in the Crater of Kilauea	80
Kate Field	98
Cottage in Dr. McGrew's Grounds where Miss Field died	102
Boki, Ruler of Oahu in 1820, and Liliha his Wife	104
Captain and Owner of the Coronet . . .	134
Map of Japan showing Track of Total Solar Eclipse	158
The Coronet dressed for the Fourth of July, Yokohama Harbor, the Olympia at the right	180
A "Float" in Matsuri Procession at Kyoto .	206
Stone Lanterns and Cryptomerias at Nara . .	212
Temple at Nara	214
View on the Railway near Morioka . . .	230
Landing the Emperor's Portrait at Esashi .	238
The Great Tidal Wave as portrayed in a Native Magazine	248
Ainu Couple, the Woman wearing Ceremonial Beads	260
A Typical Ainu	268
The Electric Commutator	278

Japanese Carpenter making Plate-Holders at Eclipse Station	280
Fanciful Lamp-Post and Native Inn at Esashi *facing*	282
Ainu holding Mustache-Lifter, about to drink Sake	290
Old Ainu Chieftain	300
Ainu Woman carrying Child and Burden	304
Articles gathered in Ainu Houses	312
Lighthouse on the Beach at Esashi (from a Drawing by Mr. Thompson)	322
Expedition Members, and Old Schoolhouse, after the Eclipse	324
A "Hairy Ainu"	340
Route of the Expedition, and Coronet's Course	354
Ainu Woman weaving Elm-Fibre into Cloth	360
Articles of Ainu Manufacture	374

INTRODUCTORY

CHASING eclipses, always of interest in itself whether the eclipse be caught or not, yields great wealth to science when these elusive phenomena are properly overtaken.

Sun and moon are of apparently the same size, and by a happy working of the celestial mechanism the moon sometimes comes directly between us and our central luminary, causing a total eclipse of the sun. But this happy state of things can by no means last longer than eight minutes. Usually in far less time sun and moon seem to slip past one another, and though for two hours or more the partial phases may continue, the duration of entire darkness is, on an average, not much over three minutes.

The astronomer wishes totality could last three hours or three months, that by the beneficent shielding of the sun's intense brightness he might have an opportunity of studying without interruption that most beautiful and mysterious sight in nature, — the outflashing radiance of the corona.

This spectacle, so impressive as hardly to admit of description, has, thus far in the history of science, been visible only during a total eclipse. Possibly in part an atmosphere of the sun, holding the secret of solar constitution and energy, what wonder that the enthusiastic specialist longs to interrogate its reticent streamers until hidden things shall come forth to his questioning telescope and camera. If a permanent eclipse would only disclose coronal secrets, any serious interference with mundane matters would give him small concern.

By a series of saddening calculations, based upon the number of eclipses in a century, the length of total phase, probability of cloud, and average number of observers and telescopes likely to address specific questions to the sun at the time of his temporary retirement; considering, too, the fondness of eclipse tracks for oceans, deserts, tropic marshes, impassable forests, and other localities where no civilized human being, not even an astronomer, can follow, the ardent pursuer reaches depressing conclusions. A miserly century, despite its seventy total eclipses of the sun, allows only about one solid day's watching of the corona. Very natural, then, the impatience to follow, if this fascinating shadow beckons toward regions even remotely accessible; and no less the desire to invent something whereby the precious three minutes, rich with tantalizing stores of coronal wealth, may virtually be lengthened many fold. To accomplish this end was mainly the object of our Amherst expedition.

Those who follow in the train of an astronomer, belonging strictly to his family, scarcely know, amid a multitude of original diversions, where to find themselves from day to day, in an existence successfully robbed of monotony. Not only does he rise at all hours after midnight and remain awake at all hours before; not only does he fill the house with developed and undeveloped photographic plates of stars and meteors, ghostly nebulæ and flying comets, as well as sketches of sun-spots and blue prints of strange apparatus; not only do piles of student examination papers, covered with frenzied diagrams, hide beneath apparently innocent magazines; and proof-sheets of forthcoming volumes lie in wait in every drawer; but one should never be amazed to meet the Professor himself at a moment's notice in any portion of the globe.

Eclipse shadows rarely fall upon him comfortably ensconced in his home observatory. Should he experience the good fortune of witnessing a single one from his domestic dome, about three hundred and fifty years must elapse before another would pass that way. Eclipse astronomers are necessarily cosmopolitan.

But, apparently erratic, these paths of darkness, like all celestial movements, are subject to definite law, only of such immense scope that one generation is not long enough to observe an appreciable fraction of its operation. In a single astronomer's lifetime, eclipse tracks may seem to obey their own sweet will, — falling, for instance, in his youth upon France (1842) and Sweden (1851), crossing Peru (1858), Spain and the Pacific coast of North America (1860) in his manhood; if still enthusiastic he would have gone to the Malay peninsula (1868), even extending his research to the great American eclipse (1878); had he begun at an especially tender age with the French eclipse, he might have retained eyesight and energy enough to journey to Egypt for a glimpse of its traditional darkness (1882). Here is variety of locality enough to confuse all theories of eclipse visitation based upon individual experience. Unhappily he cannot seek farther transits of Venus, because the next one occurs A. D. 2004, a date somewhat in advance of even the most hopeful astronomer who has the misfortune of being already alive. But eclipses and other spectacles in the firmament generally deny their beauties and revelations to strictly civilized centres. And so, although here to-day, he is to-morrow on the high road or the high seas, bound for Alaska or Pike's Peak, West Africa or the Marquesas Islands, Egypt or Chile. He speaks of these somewhat unusual localities with a familiarity not known to the tourist, and born of close acquaintance and superior companionship. He casually mentions residence for a time in Nova Zembla or Vladivostok as too much a matter of course even for comment. Truly, intimacy with immeasurable stretches of infinite space induces a just estimate of the meagre dimensions of our own planet, where few regions are impossibly remote or forbidding, if only some celestial performance be visible from their all but inaccessible wastes.

Rarely were such expeditions undertaken until the middle of the present century, and it was not many years ago that "darkening of the sun at noonday" meant unreasoning terror, even despair, to all beholders. Even now, in parts of China and India, superstitious ceremonies are performed while the "great monster" calmly devours the friendly sun. In Japan, until recently, ignorant peasants covered their wells during eclipses to prevent poison from falling into them from the sky.

If, in olden time, an eclipse occurred within convenient distance, astronomers observed the time of beginning and ending, sometimes noting the fact that a pale halo of light seemed to encircle the dark body of the moon. That this corona had definite structure, or that it offered important problems, apparently never occurred to early observers. Scientific study of the corona and of the sun's constitution has been wholly contained in the last sixty years, the significance of total eclipses of the sun being a purely modern recognition.

Although invented in 1839, photography was first successfully applied to a total eclipse of the sun in 1851 at Königsberg, securing a fine record, not only of the wonderful red prominences which burst forth at totality, but of the mysterious radiance of the ethereal corona. Henceforward, advance in this method of observation was rapid, and in 1868 Janssen in India, after the longest eclipse ever observed (about five minutes and a half), announced the epoch-making discovery, that the protuberances can be studied by the spectroscope without an eclipse, — that is, in full sunlight. Yet the character of the blood-red jets is not in all respects the same as when the moon's dark body makes the screen, so that necessity for continued research upon them during eclipses still remains. But no triumphant observer has yet reported success in seeing the corona without an eclipse, though many trials with highly sensitive instruments have been made. Future years may bring, too, this longed-for achievement.

Prior to 1860 it was not even certain that the corona belonged to the sun at all. The outer streamers, sometimes extending ten or eleven million miles into space, were discovered by Professor Langley in 1878, from the summit of Pike's Peak. A material found in the corona by Professor Young in 1869 was named "coronium," being unlike anything known upon earth; and his marvelous "reversing layer" — when for a second or two before totality all the dark lines in the spectrum suddenly flash forth in great brilliance (seen for the first time in Spain in 1870, and confirmed in 1874 at Cape Colony) — was photographed during the eclipse of 1896 by Sir Baden Powell's party in Nova Zembla, and abundantly verified in India two years later.

Thus, bit by bit, our stores of knowledge of the corona accumulate. Finely equipped expeditions to follow the fleeting pathway of shadow are in our day constantly sent out, often by government, and the leading nations of the world vie with one another in the amount of valuable material gathered by their astronomers.

Questions of probable cloud are, of course, very important. As the shadow will fall over localities known many years in advance, observations of prevailing sky conditions are possible during several seasons beforehand all along the track of anticipated darkness, and from a comparison of them, regions least cloudy can be chosen.

Selecting the site for an observing station involves great, almost terrifying, responsibility. Of four or five available places, one may prove clear and the others cloudy on the fateful day; or one may be overcast while the remainder rejoice in brightest weather. And the wrong one may have been chosen.

Aggregation of many observers in one region is less desirable, scientifically considered, than various parties scattered along its line; for possibilities of cloud-interference are less, and it is desirable also to know whether the corona during a single eclipse presents exactly the same features to eyes hundreds of miles apart. In other words, whether or not it may change in two or three hours. As the track is ordinarily many thousand miles in length, this scattering of observers along the land-line is by no means impracticable.

The eclipse of 1896 offered a variety of opportunities. Beginning in Norway, the track lay across frozen Nova Zembla, through Siberia and the Amur River region; thence crossing the Sea of Japan it traversed the Hokkaido, or northern islands of the Japanese Empire, losing itself at last in the Pacific Ocean.

Norway dismissed from consideration as our goal because several other parties of observers had planned to locate there, Nova Zembla was investigated; but the eclipse not occurring until the ninth of August, and the Coronet having no steam, it was deemed inexpedient to remain so late in the far north. The prospect of a possible winter ice-bound in an Arctic harbor was not sufficiently alluring to risk the reality.

For three years, at Professor Todd's request, meteorological observations had been made in Japan, throughout the region of coming eclipse; and Yezo, the largest northern island, was made the destination of the Amherst Eclipse Expedition.

In October, 1895, plans were laid, instruments and their mountings begun, and the Coronet was preparing for her long voyage around the Horn.

DEEP-SEA YACHTING

BY ARTHUR CURTISS JAMES

To the yachtsman truly interested in his hobby, who enjoys a home on the rolling deep for its own sake, deep-sea cruising affords a wider scope and more perfect enjoyment than can possibly be obtained from short trips on inland waters.

The coast of Maine and the waters of Long Island Sound are unsurpassed anywhere in the world as headquarters for Corinthian sailors, but it is not until "Farewell" has been taken and the first course set for a distant port, that the true lover of the sea begins to feel the exhilaration of life on the ocean wave. Newspapers are not wanted. Telegrams are impossible. Worry is left behind, and the yachtsman enters upon an indefinite period of perfect contentment. Details of managing the vessel, the study and practice of navigation and seamanship, even settling the quarrels of sailors and cooks, are simply pleasant pastimes. Events which on shore would cause endless annoyance and trouble, at sea mean simply more work and wider experience. Storms, fog, accidents, are to the sailor only incidents, and every new difficulty arising suggests a way to meet it.

The yachtsman who is able to do so should command his ship at all times, and particularly on a long ocean voyage. He will find more opportunities to improve his navigation and to develop seamanlike qualities in one month at sea than in three years' regular yachting near home. He should be thoroughly familiar with every department of his ship, and by so doing he may rest assured that time, even on the longest voyage, will not hang heavily on his hands.

Probably every one who has been to sea has a different theory as to the best class of yacht for a long ocean voyage. Designers have given us everything, from the immense floating steel shell inclosing five thousand horse-power, to the able little pilot boats remodeled with all the comforts of a yacht. The Coronet is practically of the latter class. Built in 1885, of one hundred and fifty-two tons net register, one hundred and thirty-three feet over all, twenty-seven feet beam and twelve and one half feet draught, she has since that time covered a greater number of miles than any other American yacht. Her career was opened, and her first reputation made, by defeating the famous schooner yacht Dauntless in a midwinter race from New York to Queenstown. Shortly after, she rounded Cape

Horn during the worst season of the year, making the voyage from New York to San Francisco in one hundred and five days. Her trip around the world, completed in thirteen months, was followed by four trips to Europe, and two to the West Indies.

The completion of the Japan trip has added forty-five thousand miles to her record. During her whole history, she has never lost so much as a bucket from her decks, nor met with any serious mishap. From my experience on the Coronet, I should not know how to improve upon her for a strictly sailing deep-sea cruising yacht, dry and comfortable in all weathers, and able to keep the sea and make passages with almost a steamer's regularity.

There is an old sailors' maxim that "they who go down to the sea in *ships* behold the wonders of the deep, but they who go down to the sea in *schooners* see Hell," and without doubt this saying has considerable foundation in fact. For running before a light wind in a heavy following sea, the long main boom of a large schooner yacht is certainly a dangerous companion; and should it break loose, it would be likely to take charge of the deck, and almost certainly cause serious damage. For ocean work a squaresail is an absolute necessity, for then the foresail and squaresail can be set and the mainsail taken in when running. The squaresail yard also gives scope to one's ingenuity in planning additional skysails and other forms of balloon canvas. The utility of this rig was soon discovered after sailing from San Francisco.

The coast of California is by no means an attractive place for yachting. The glorious climate for which the state is so renowned confines itself strictly to the land, while at sea fogs, gales, and calms alternate with surprising regularity.

It was without much regret that we left our anchorage in San Francisco Bay, and headed the Coronet for the Hawaiian Islands. The San Francisco bar demanded tribute from most of the party, but at length farewell was taken at the Farallones, and the trip to Japan fairly started. The first week's run was the poorest in the history of the vessel, averaging only one hundred miles a day; but after finding the trade winds in latitude 27° N., the Coronet seemed ashamed of herself, and made two hundred and fifty miles daily in the effort to retrieve her reputation. A considerable part of the distance was covered without aid of the mainsail, under foresail and squaresail.

The evening of the fifteenth day found us safely anchored in the snug little harbor of Honolulu. It is not the province of this chapter to describe the beauties of the Islands, nor to dwell on the delights of a visit to what has been well named the "Paradise of the Pacific," and could with equal truth be called the Paradise of the World. Even after extended acquaintance with the undeniable and oft described charms of "Picturesque and Progressive Japan," it is enough to say that our stay in the Islands was the most delightful of the entire trip; and it was the unanimous hope that the mother country might become better acquainted, and more closely united to our countrymen of the Hawaiian Republic.

To the yachtsman, the islands of the Pacific lying north and south of the equator afford an inexhaustible field for most delightful cruises. From the latitude of Honolulu south to Pitcairn and stretching across the Pacific to Australia, are thousands of islands, many of them inhabited by curious and interesting races.

Yachtsmen have been criticised, and in some cases justly, for using their magnificent fleet of vessels as mere toys. What an assistance they might be in advancing our knowledge of geography, if their pleasure trips could be turned to some practical account! With plenty of time, which is of course essential to a thorough enjoyment of any cruise, and with a properly equipped yacht at one's command, I know of no part of the world which would better repay a visit, or which could yield more valuable results in extending geographical and commercial knowledge, presenting, as it does, so wide and unexplored a field for scientific research.

Our knowledge of the islands of the Pacific is at best exceedingly meagre, and there is certainly no class of men better fitted, either by education or equipment, to coöperate with the navy in adding to our store of information than the yachtsmen of the United States.

The object of the expedition made it impossible for us to linger long in Honolulu, and inadvisable to make any other stop; but if the fates favor, the Coronet may before long again be headed for "The Paradise of the Pacific" and the islands of the southern seas.

There is no need of waiting for a fair wind or favorable weather to start on a cruise from the Islands. The trade winds are practically always fair, and the sailor need seldom look for anything more terrible than a rain squall to interfere with his plans.

The afternoon of the 25th of May found the Coronet again in her element, out of sight of land, with boats lashed securely on deck and everything snug below and aloft; prepared for anything which might be in store for her on the four thousand miles of sea that must be covered before reaching Yokohama. The sailing course from Honolulu to Japan is considerably longer than that followed by the steamers, but the time at sea might have been doubled and still no one would have objected, so delightful was the entire trip. In order to hold the trade winds, we kept between the parallels of 18° and 20° north latitude almost the entire distance; and so perfect summer weather was assured. Day after day the awning was set on the quarterdeck and the yacht kept on her way with scarcely more motion than would be experienced in Long Island Sound. The long Pacific rollers lazily following, and even the flocks of goonies slowly circling astern, seemed to express the spirit of the tropics and bid us enjoy southern seas to the utmost. Although not strong, the trades were almost absolutely steady, and gave us an average of about one hundred and fifty miles a day for the trip.

During the typhoon season the coast of Japan is not a particularly inviting place for vessels of any class, and when our log showed that we were about two hundred miles from Yokohama, the barometer beginning to fall rapidly with constantly increasing wind and a heavy sea, we thought it time to prepare for a warm reception to the country. Evidently we were on the edge of a revolving storm, the centre of which appeared to be traveling rapidly along the coast. Under short sail the Coronet was kept on her course until nightfall, but the constantly and rapidly falling barometer warned us that it would be unwise to attempt to approach land until the disturbance had passed. A storm at sea may be a grand sight, but a little of it goes a long way, and the grandeur of the fury of the elements did not compensate for the prospect of being hove to for three or four days within a hundred and fifty miles of port. During the middle watch the gale moderated, and at dawn we were able again to make our course. The passing of the storm, however, had left behind it a very heavy sea which delayed our progress, and it was nearly midnight of Sunday, the 21st of June, when the light on Mila Head which marks the entrance of Yeddo Bay was sighted.

Yokohama pilots are an unknown quantity. No response came to our repeated signals, and we were obliged to navigate the channel unaided. During the night we had our first introduction to the methods of navigation employed by Japanese fishermen. They sail their unwieldy junks without lights and without the slightest regard for the "rules of the road." Their immense square-sail is an impenetrable wall between the helmsman and anything which may be ahead of him. A lookout is an unheard-of precaution, so it was only by rare good fortune that we avoided running down a number of them in the darkness.

By ten o'clock on the morning of the 22d we had covered the forty miles between Mila Head and the breakwater which forms the harbor of Yokohama, passed the quarantine officials, and dropped anchor close to the magnificent United States cruiser Olympia.

One of the most delightful experiences to the yachtsman on summer cruises in home waters is the harbor life in such ports as Bar Harbor, Newport, and the other resorts of our eastern coast. To many this social life is the highest ideal of yachting, and were it eliminated, the chief charm of the sport would be taken away. In foreign ports such experiences are by no means lacking, and are on the contrary far more interesting and attractive than at home.

In such a country as Japan the government is most friendly to Americans, and an American yacht receives courtesies equal in almost every respect to those granted to men-of-war. The constant interchange of civilities with the officials of a country whose manners and customs are so entirely different from our own is a source of never failing interest, and the yachtsman's welcome to the local yacht clubs of Oriental ports is more hearty and sincere than seems to be bestowed by nations which make greater claims to yachting fame.

Yokohama is a favorite rendezvous for the

ships on the Asiatic station during the summer months, and the most delightful memories of the entire cruise are the friendships among the officers of the Olympia, the Detroit, the Yorktown, and other ships of the squadron. The time has passed when an American need blush for his country on meeting our naval vessels abroad. The ships that carry the stars and stripes in the Asiatic squadron are second to none, and the officers are worthy successors to those who in early days made American seamen famous the world over.

Opportunities for cruising along the coast of Japan are very limited. Particularly in summer, the danger of typhoons and the absence of available harbors make it unsafe to take extended cruises. A trip through the Inland Sea, however, is one which can safely be taken by any yacht, and which no yachtsman visiting Japan should miss. Owing to exceedingly poor transportation, this remarkable combination of land and sea has not received the attention it deserves from writers on Japan. Among the Japanese, it is considered one of the three principal sights of the country. The steamers of the Pacific Mail and other lines sail through a part of the Sea on their regular trips, but the main ship channel gives no idea of the quaint little harbors, charming scenery, and interesting out-of-the-way places which can be visited, for the present at least only by a yacht or specially chartered steamer. For a steam yacht there are no difficulties of navigation to be overcome, and all that is necessary is to obtain a pilot thoroughly familiar with all parts of the Sea. A sailing yacht, however, requires the constant attendance of a tug in order to pass through the most beautiful, but exceedingly narrow passages between the islands. Even with such assistance, a sailing vessel should not attempt to pass the narrowest straits, except at slack water. Many passages are less than one hundred yards wide, through which the tide rushes at the rate of ten knots and more.

Picturesque and perfectly sheltered harbors are numerous. Some of the ports at which we stopped had never been visited before by foreigners, and the little remote fishing villages afforded a splendid opportunity for studying Japanese character, untouched by Western civilization. The ten days spent in the Inland Sea were altogether too short a time to explore its intricate channels, and even to sail past the thousand mountains and thickly wooded islands which form a barrier to the Pacific and give the Sea its name.

It is difficult to believe that the ocean north of the fortieth parallel is the same old Pacific over which we so peacefully sailed from Honolulu to Yokohama. In the south the principal occupation was endeavoring to devise new balloon sails to catch every breath of the light trades, while on the return trip it was frequently a scramble to lower all sail in the shortest possible time. Instead of balloon canvas, the thought was to see how small a rag could be shown to the gales. The yachtsman who wishes to enlarge his experience and desires practice in handling his vessel under all conditions of wind and weather should cruise in the Pacific Ocean.

Leaving Yokohama the 2d of September took us to sea at the worst time of year. On the day before sailing a severe typhoon had passed up the coast, and three days later we encountered the edge of another which did immense damage about two hundred miles northwest of our position. It was, then, with a feeling of relief that we found ourselves at the end of a week's sailing beyond the reach of such unwelcome visitors.

From a study of the chart, one is led to expect a current setting along the coast of Japan and across the Pacific far greater in volume and strength than the Gulf Stream of the North Atlantic. The *kurosiwa* or Japan Current undoubtedly exists, but it would seem to be far more frequently affected by the prevailing winds than is the Gulf Stream. Directly in the supposed centre of the stream where a current of from one to three knots an hour was expected, we were surprised to find by observation that practically no help was received from this source. Absence of this current was still more of a puzzle as we had experienced only westerly and southwesterly winds, which should have increased rather than retarded its force. The only plausible explanation to account for temporary cessation of the Japan stream is that the typhoons which had been very numerous during the month of August had, on leaving the coast of Japan, become strong northeast gales. This theory was strengthened by our meeting a heavy northeast swell lasting until after we had passed the 180th meridian. It had been our purpose on leaving port to follow as closely as possible the great circle track to San Francisco, and we were fortunate in being able to make practically a perfect course the entire distance.

There was certainly no monotony in the sailing. Frequently a whole sail breeze would begin the day, increasing by night to a howling gale, followed by a few hours of flat calm. In order to realize our hope of making a reasonably rapid trip, constant watching and active work on the part of all hands were necessary, so that the short and precious hours when it was possible to drive the Coronet to her utmost should not be wasted.

One gloomy, breezy morning, an immense waterspout appeared less than two miles from us, traveling toward the northwest. It was a grand sight, but not a pleasant neighbor, and no one regretted its final disappearance astern.

As we approached the coast of California, fog, the sailor's worst enemy, shut in upon us. For three days observations had been impossible, and we were obliged to rely upon dead reckoning, which although always kept with great care, at this time received double attention.

At eight o'clock on the evening of the 1st of October, we judged our position to be about ten miles to the westward of the Farallones lighthouse; and as the fog continued dense, hove to, waiting for a more favorable chance to run for the light. At midnight the fog "scaled up" a little, and the Coronet was headed a true east course. Scarcely an hour passed after getting under way before we heard a whistle right ahead, which soon proved to be the siren on the Farallones.

The yachtsman who has never known the pleasure of making a light after a long and difficult voyage has something to live for. Even the professional seaman knows the exhilaration of the moment, and the amateur may be pardoned if he too feels a thrill of pride and pleasure. The wonders of the universe never seem so close and real, as after a month at sea with nothing but the sun and stars to mark one's path. By their help we made within fifty miles of the shortest possible course between Yokohama and San Francisco, covering the forty-six hundred miles in thirty days.

After passing the light, fog settled again, and the anchorage off San Francisco was reached by aid of the numerous fog signals along the shores of the Bay, after having caught only one glorious glimpse of the Golden Gate.

In concluding this chapter on the strictly deep-sea cruising of the Coronet, I cannot refrain from urging yachtsmen in general, and those taking ocean trips in particular, to coöperate with the Hydrographic Office in adding to our knowledge of ocean currents, winds, and other phenomena of the sea. Foreign nations recognize our Hydrographic Office as a model for all countries, and its high standard of excellence can only be maintained by the hearty assistance of all interested in seafaring matters. The information which it furnishes to mariners is of the greatest value, and the daily observations upon which this information is founded can easily be taken on any properly equipped vessel. Our government is most generous in its treatment of yachtsmen, and it seems only proper that we should do everything in our power when opportunity offers to assist in placing the maritime affairs of the nation on a basis truly representative of American thought and American progress.

CORONA AND CORONET

CHAPTER I

THE CORONET

Swift flies the schooner careering beyond o'er the blue;
Faint shows the furrow she leaves as she cleaves lightly through;
Gay gleams the fluttering flag at her delicate mast —
Full swell the sails with the wind that is following fast.
                                        CELIA THAXTER.

YEARS ago, a prevalent style of tale possessed never-failing interest, though causing continual surprise to one small reader. Ordinarily the work of English authors, some boy-hero was frequently despatched to India, usually because of sudden poverty or other disaster overtaking his relatives; and the impression given was that, next to death, a journey to the antipodes was the most dismal of fates.

While accepting the story-teller's point of view so far as necessary in sympathizing with the sorrows of the leading family, I was always filled with amazement that a journey to India could be regarded as a calamity. I half wished I might have been that youth setting off to seek his fortune in far lands; perhaps a faint foreshadowing of a later time, when I should become an adjunct to the family of an astronomer, one of whose specialties should be interrogating a hidden sun.

Whatever the reason, strange journeys to remote regions have always meant delight, and had time been plenty, the peerless Coronet might have had a passenger on her trip around the Horn, instead of awaiting the entire party at San Francisco after this portion of her cruise was over.

Designed in large part by Captain Crosby, for many years her sailing-master, as well as by Messrs. Smith and Terry, she was built in 1885 by C. and R. Poillon, of Brooklyn, at a cost of about $70,000. At the time of the Japanese expedition the largest sailing yacht in the New York Yacht Club, her finest record is in two consecutive watches of sixty miles each, thus accomplishing 120 miles in eight hours.

Although the actual dimensions of the yacht are given by her owner and captain, with a few words as to her history, he has not described her beauty, the elegance of her interior arrangement, and the details of the race that opened her famous career so brilliantly. The start was from an imaginary line off Owl's Head, Long Island, at 1.10 P.M. of the 12th of March, 1886, the finish off Roche's Point, Cork, Ireland. The Coronet occupied 14 days, 19 hours, 3 minutes, and 14 seconds in the passage, winning the race by 1 day, 6 hours, 39 minutes, and 40 seconds, sailing 2905 miles; while the Dauntless sailed 2957 miles, — a fine race, always spoken of as "a glorious victory, an honorable defeat."

Immediately afterward her owner made a voyage around the world, the graceful yacht exciting much admiration in all ports. At Honolulu, King Kalakaua came on board, and in Yokohama harbor she was visited by the Emperor, who ordered at once for himself a boat exactly like the Coronet's gig.

In October of 1893, she became the property of Mr. D. Willis James and his son.

She is white, schooner-rigged, carrying every sort of sail, and as airy as a bird. It is not to be expected that any wandering breeze, however light, could escape all her alluring opportunities for usefulness in topsails, staysails, jibs, and raffies, — and, indeed, when this cloud of canvas is spread to a brisk wind, the Coronet is a thing of beauty indescribable.

With gig and cutter stowed away forward for a long voyage, a fine stretch of open deck still remains, while, no room being wasted on engines or coal bunkers below, all the space is available for living quarters. Finished in carved mahogany, the main saloon is about twenty feet square. A piano and writing-desk, easy chairs and divans invite varying moods, bookcases are filled with tempting volumes, and an open stove of red tiles shows a glowing bed of coals in damp or chilly weather.

Two large staterooms, also finished in mahogany, contain brass beds, furniture and walls of one done in pink velvet, the other in satin brocade. With four other rooms, each artistically furnished, ten or twelve guests are luxuriously accommodated.

A crew of ten men, a sailing-master and two mates, a cook with two assistants, and two stewards, the Coronet's freight of human beings on many trips falls little short of thirty.

As she lay during the autumn of 1895 in Tebo's Basin, South Brooklyn, all her possibilities of beauty, speed, and grace latent, preparation for her long voyage around the Horn went rapidly forward. Rigid examination revealed a tiny spot in the huge foremast. The imperfection, less than an inch in diameter, hardly made an indentation on the surface of this great timber, yet at some crucial moment a sudden strain might come upon just that spot. So a new and flawless mast was substituted. No less minutely was inspection made of the whole vessel. New steel rigging was provided, a thick coat of paint covered the white deck for the voyage to San Francisco, furniture was shrouded in linen, and heavier parts of eclipse apparatus already complete were carefully stowed below.

On the 5th of December, 1895, she left her cosy winter quarters to breast the icy seas and gales of a four months' voyage. In the southern hemisphere summer weather would prevail, but many days lay between the Narrows and that genial region.

Her owner and his wife, the Astronomer and a few guests went down the harbor on the yacht, and, returning with the pilot, watched her lightly skimming the wintry waters farther and farther from sight, as early December twilight settled over the tossing sea. Great faith is required in the science of navigation, in the seaworthiness of his craft, and the skill of his sailors, for a yachtsman to entrust his dainty vessel to the mercy of winds and waves during a voyage of fifteen thousand miles.

Five days later the Coronet was sighted by the steamship Bræmer, nearly a thousand miles from Sandy Hook, encountering heavy seas upon the edge of a severe storm through which the Bræmer herself had come. Occasionally other vessels were sighted, but they were not bound in directions for bringing news — and this was the only report during the long voyage. So the winter was passed, with reasonable certainty, but no knowledge, that she was making her course safely.

Considering the Coronet's sailing qualities and former achievements, this did not require an impossible exercise of philosophy. Once each week her owner plotted her probable course and run upon the chart, his faith supplying deficiencies in actual news.

The Coronet's log during all these days is an interesting record. Many fairly good runs are set down, but she encountered much rough weather, frequently a "whole sail" breeze; and suggestive, even if painfully succinct accounts are given of the various sorts of weather, vessels sighted, gales coming on, guns taken below, all sails reefed, and "oil-bags got ready."

For Sunday, 9th February, 1896, off the coast of Patagonia, the entry reads: "At midnight wore ship on account of the sea. Ship burying herself to the foremast, middle part. Called all hands and reefed her down fore and aft, and wore ship. Latter part much rain and blowing hard in squalls."

Farther on are records of "Confused sea. Rain. Hove to under the fore trysail. Got the oil-bags over side, one from each cat-head, and one in the main rigging."

On Thursday, 13th February, 1896, "Blowing strong. Lying under reefed storm sails, and oil-bags over the side, and an old Cape Horn swell running. It seems as if the little Coronet would go end over end at times. But up to today we have not lost a rope-yarn off the deck."

Two days before anchoring in the harbor of San Francisco, a high, confused sea was still running, and "a good deal of tumbling aboard." But she soon sailed triumphantly into port, casting anchor at Sausalito, headquarters of the local yacht club.

CHAPTER II

PREPARATION

Pause not to dream of the future before us.
OSGOOD.

PROFESSORS of practical astronomy must always invent. No mental graces or acquirements can supersede a mechanical bent, whereby instruments of whatever sort give joy and all telescopes delight, merely in themselves, and quite independently of their performance in bringing heavenly bodies a few million miles nearer.

Since in this generation we cannot make sun and moon stand still, lengthening of the precious minutes of totality can be accomplished only in two ways. One astronomer might take with him ninety-nine others, each with telescope, camera, spectroscope, or other bit of apparatus to ask his own particular question of the calm corona as it gleams against the silent darkness. Or one astronomer could transport a hundred telescopes and cameras, if only each could make its own record. In the history of science thus far, eclipse expeditions of one hundred human observers have not materialized, although an attractive prospect to regions unheard of where such a mission might establish itself. But a composite machine is possible, by which a hundred instruments are able to ask simultaneously a hundred different questions of the corona automatically, while one astronomer sets everything in motion, placing safe and implicit reliance in the precision of his mechanism. Fortunately, too, machinery has no nerves; for in the past, impressiveness of the scene at totality has been responsible for many a lapse in executing well-rehearsed programmes.

While the Coronet was buffeting Cape Horn swells and the great rollers of the Pacific, carrying tubes and mountings, the Astronomer was hard at work completing his invention in finer detail, until satisfied that the prospective minutes of total eclipse would be lengthened at least tenfold. Specifically, twenty telescopes and cameras were to observe and set down at the same time, all under electric supervision of one central mechanism; and exact records of the unemotional tool would be at hand after the eclipse was over, well adapted to patient study at leisure.

So who could complain if tubes and valves and pneumatic arrangements and object-glasses and electric devices of every sort strewed the drawing-room, and measured their innocent length on every floor throughout the house? The family of a professor of astronomy get thoroughly accustomed to all such trifles, and learn to step circumspectly among polished brass and shining specula, nor can they by any chance be surprised at strange occupants of their desks and dressing-tables.

The cardinal principle of this automatic device is simple enough even for comprehension by the unmechanical, as an astronomer's relatives are too apt to be. Research on the corona has become in these latter years mainly photographic; so that a multitude of telescopes and spectroscopes, if transformed virtually into cameras, are thus able to collect their evidence simultaneously and independently.

The idea that machinery could be made to execute these motions, instead of separate persons, first occurred to Professor Todd during his former expedition to Japan in 1887. The plan was roughly carried out by native assistants at Shirakawa, on the old castle donated by the Government for his observing station. Although crude, the separate mechanisms worked so well that he developed the same theory more elaborately for his next expedition — to West Africa in 1889. Exhaust air currents through pneumatic tubes, connected with each telescope and plate-holder, were controlled by a slowly moving perforated sheet of paper, similar to those now familiar in

EXPEDITION HEADQUARTERS AT ESASHI, SHOWING PORTABLE HOUSE AND TWENTY TELESCOPES

automatic musical instruments. Movements of absolute precision allowed the exposure of over three hundred plates during the period of totality.

The apparatus proving somewhat bulky, electricity was made the controlling power when in 1895 plans were maturing for Yezo. Endless chains of plate-holders of different sizes were arranged to pass before each of the twenty telescopes, at varying rates of speed. A copper cylinder full of pins revolved slowly, each pin as it passed along touching its appropriate metal tooth, and closing a circuit that set in motion some particular instrument at any prearranged instant during the two minutes and forty seconds while totality should prevail. A moderate calculation of its capacity showed that four hundred pictures could be taken, the movement of each exposing shutter and its corresponding plate-chain being deliberate and precise.

That all these telescopes should remain constantly pointed at the sun, even for two or three minutes on this slowly whirling earth, farther mechanism was necessary. Without a driving-clock of some sort, any celestial object is speedily out of the field of view, or off the plate. First, all the telescopic cameras are rigidly attached to one central frame, and this polar axis must itself follow the sun in his apparent path through the sky, carrying with it all the instruments. A sand-clock, used successfully in West Africa, was thought again feasible. By this arrangement a heavy weight resting upon a tube of sand slides gently down, as the sand runs out below at a uniform rate, hour-glass fashion.

After duly experimenting, the Professor decided reluctantly that the sand was not, after all, sufficiently smooth for his purpose. Ultimately a column of glycerine was substituted, to his entire satisfaction.

Preparation for an attractive expedition has one curious phase, — the variety of demands to join it, a few delivered verbally, though chiefly by letter. From every walk of life and all parts of the country came insistent applications for billets, possible and impossible; each setting forth in glowing terms the writer's especial qualifications. Every mail for many weeks brought such letters, — a unique collection.

Certain aspects of scientific expeditions, too, are not represented by clocks and lenses, nor the critical selection of personnel. Apparatus did not contain the whole winter's story. Seven months' absence from one's native land means many costumes. The rigor of our own northern regions, and of the first days on the Pacific, the tropic heat of Honolulu and Yokohama, the memory of Japanese humidity (gloves had been sealed in Mason's jars to prevent moulding), camping-out gowns for the eclipse station, full dress for all kinds of entertaining in foreign and always jovial ports — to provide for all these contingencies may not have necessitated the quality of brain for inventing twenty electric observers of an eclipse; but the problem was not entirely simple, nor was the time too long to prepare for conditions so varied. By the middle of March, a focus was approaching. Tests of apparatus were nearly complete. Crates and trunks and boxes were in readiness; and one hundred and seventeen days had passed since the Coronet left New York. At last, on the first of April, the welcome telegram arrived — "Coronet in San Francisco to-day. Can you start to-morrow?"

The first stage of expedition travels began as early bluebirds were singing their blithe spring songs among the budding trees of the old college town, and a long good-by was said to its classic groves.

CHAPTER III

OVERLAND

> Be Mercury, set feathers to thy heels,
> And fly like thought.
> SHAKESPEARE, *King John*, iv. 2.

A RADIANT Easter Sunday was followed by heavy snow, submerging New York, when friends collected at the Grand Central station to say farewell that early Monday morning. Despite gloomy skies roses filled our hands, the College Glee Club gave the Amherst yell, and the long journey began, with its sense of exquisite rest and lack of responsibility after constant and fatiguing preparation. Quick or careless movements, however, were indulged in with caution, from consciousness of our precious surroundings, — lenses, chronometers, photographic plates *ad libitum*.

At Rochester more expedition material appeared; and continually we were met, not only by friends and well-wishers along the route, but by waiting instruments. Awakened in the night by a stop — arousing thought immediately concentrated upon "another telescope!"

Mr. Hill, president of the Great Northern

road, had generously reserved for our use his own private car (" A 1 "), which at Chicago was quickly filled with expedition possessions, and the various members of the party sallied forth for a day in the city. With the fall of windy twilight more farewells, and pleasant last words from the president of the World's Fair Commission, and the discoverer of the fifth satellite of Jupiter.

There had been days of well-remembered plain in crossing the continent by the Canadian Pacific years before, but the mental effect was somehow different from the impressive and illimitable levels of North Dakota and Montana. Strange to weirdness and unutterably lonely, snow often fell across the treeless wastes, no trace of spring brightened the gray scene, and twilights descended in ghostly fashion, as the edge of the visible world softly faded.

Much of the landscape was merely clay, sometimes low, but menacing hills and ridges, fantastic, waterworn, — miniature Gardens of the Gods done in mud. Here and there paths and tracks led to nothingness. In the Fort Peck Indian reservation spiritless communities collected for no apparent reason; log huts encircled tepees flying scarlet flags, and a brawny squaw, chopping wood with vigorous strokes, was watched with silent approval by a row of braves. Occasionally a cowboy sped along, and companies of Indians in vari-colored rags galloped about on rough ponies from nowhere to nowhere. The days were dull and cold like late November; a ray of genial sunshine might have lighted these infinite plains with almost cheerful life, making swift shadows and gleams of brightness, but under the sombre sky they were dead, impassive. And still trails wandered off aimlessly, the wind blew drearily, and the buttes or mud cliffs on the horizon held out no promise beyond their hopeless verge.

An exceedingly fine road-bed these level lands afford the Great Northern, our luxuriously appointed car riding so smoothly that letters and journals were brought surprisingly up to date, and expedition work suffered no interruption. Life went on with great cheerfulness, whatever the outer scene. It was an early discovery that the personnel of the expedition included contralto and soprano voices, and that the General and the Musician sang fine bass and tenor. With an ample supply of glees, madrigals, and anthems, many hours were spent in "reading," whose effects might not have discredited a more stationary quartette. Half unconsciously, too, the company studied one another, deciding that it was a harmonious combination as well in ways other than musical, and likely to remain so.

One memorable morning, against a royal background of blue sky, peak after peak rose into early dawn, deeply snow-covered, and inexpressibly solemn in that silent land. To fall asleep in a country of bare and limitless level, and to awake amid primeval cedars, pines, and spruces, rising straight and clear a hundred feet into blue air, and white mountains so high that their summits are invisible from car windows — how thoroughly American the contrast of consecutive days! This radiantly sunny forenoon was spent chiefly on the observation platform. Tumbling Flathead River followed for miles, and one great peak like the Matterhorn appeared and reappeared with superb effect, between the giant shoulders of nearer hills.

The Kootenai River was a clear, green stream with flashing white foam in its swifter shallows, and our train, now far above on the mountain side, perched on a high trestle, or shooting through ten tunnels, was again close beside the water, where an occasional fisherman or boatman gave accent to a landscape never lonely, though almost untouched by human influence.

Spokane was approached over level regions once more, beneath a sky like June, though great evergreen forests continued, and the snow-covered Rockies formed an edge and finish for the world. It is a sunny city, fair and attractive, and the country around was inundated with flowers, like a brilliant sea of pink and yellow and purple blossoming. Over fertile fields, miles square, where men were ploughing rich soil, the mountains retreated into the east; then train and plain were covered by a cloud, while, more ethereal in blue distance, snowy peaks caught sunlight yet, like the veritable entrance to some celestial region beyond imagining. Twilight came on softly, mountains faded, and smooth gray blotted out the world. But where sunset should have been were streaks of pale yet bright apple-green among the slate-colored clouds,— full of hope and promise. At every stop the clear pipe of early frogs filled the still evening.

Sharp contrast again with morning — we were once more among white hills, and tall evergreens straight and majestic, every branch heaped high with feathery snow. In the utter silence and dim air the falling flakes could almost be heard. Thoughtful railway officials had sidetracked our car at Cascade Tunnel over night, to await a special engine sent to take us over the "Switchback" by daylight. This pass is more than four thousand feet in elevation, and the road zigzags backward and forward until from the summit one may look down upon loop after loop below, each at a different level. Steep as were the mountain sides, yet evergreens clothed them with beauty to the very peaks, now lost in drifting whiteness. But snow covering is not permanent, nor are there glaciers, as in the British Rockies.

A little hamlet of half a dozen houses lay nine hundred feet below, with no apparent way out. Shut in on every side by steep mountains and heavy forest, Wellington's horizon is seemingly halfway zenithward.

Spring snowslides not infrequently fall across the track, when rotary snow-ploughs come to the rescue. At Wellington, word having just been brought that a train somewhere in the mountains needed release, the expedition was invited to see the "rotary" in vigorous operation, throwing ice and snow far down the gulch, and clearing the track speedily and effectively.

All the peaceful Sunday was spent at Wellington. A walk along the track in the utter solitude brought overpowering consciousness of the close immensity of those impenetrable heights. Silence was insistent. Faint murmur from a muffled brook in the valley below and an occasional bird-song, wild and sweet, drifting down into the white day from some unknown elevation, only intensified the profoundly solemn quiet.

Toward twilight the storm abated, allowing a view of the paths of former avalanches straight down mountain sides where tallest trees had been torn away like shrubs,— narrow white tracks through the forest. Approaching night filled the deep valley brimful of purple shadow; the air grew warmer, trickling streams from overhanging drifts added a sound of rushing waters. Lights flickered picturesquely from a train a few miles up the mountain, and a whistle now and then came down from the heights.

The Skykomish River escorted the expedition through breakfast, among blossoming fruit trees to the shores of lovely Puget Sound,— green water touched with white caps, and rocky shores skirted with familiar evergreens like the coast of Maine. Beneath many-shaded gray clouds the radiant Olympian Mountains shone forth fitfully, white and high, occasionally gleaming in brilliant sunshine, sparkling gates of some Paradise of Peace.

Seattle is nobly situated on successive terraces above the Sound, the Olympian and Cascade ranges in plain sight; beautiful Mount Baker and lofty Rainier. But mist and rain are overfond of hiding this unparalleled scenery. Crimson wild currant was everywhere in blossom, and the wall-flower; lawns were smoothly green, and English ivy covered many dwellings with its dignifying touch. Still unfinished, the city abounds in possibilities.

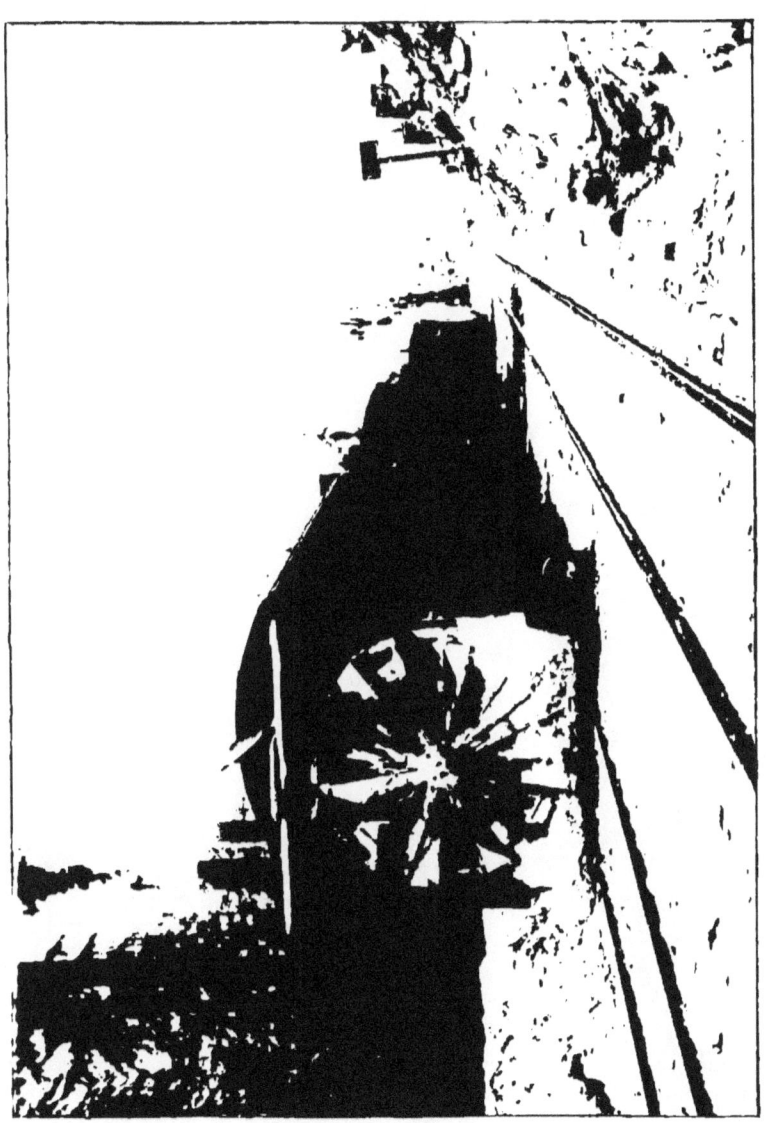

THE ROTARY SNOW-PLOUGH AT WELLINGTON

Friends were here also, and loyal Amherst graduates; but from New York to San Francisco newspaper reporters were omnipresent. Our instant arrival in every city was greeted by papers containing "full accounts" of the expedition, with ghastly portraits as well, dark and sinister, less like a peaceful body of innocent scientists than some band of outlaws bound for gore and gold. With each new stop more reporters scrambled for more material for still other "stories." But at the precise moment when pads and pencils were hopefully brought forth, the Captain, the Professor, even the amiable Doctor and General, by a series of curious coincidences, had immediately pressing business at some distant point. Others in the party seemed to melt away imperceptibly, and it so often devolved upon the present historian, deserted by her allies, to sustain the conversation on these somewhat trying occasions, that she became expert to a melancholy degree in answering questions about the plans, objects, incidents, and personnel of the party.

Often these interviews were prettily embroidered by the active imaginations reproducing them. One paper announced that the Coronet was now awaiting her guests, having just arrived at San Francisco from New York "via the Isthmus." Another stated that "Mr. James is the fourth owner of the Coronet, she having had three before him." Still another, confusing a dignified scientific expedition with a party of Dunkards simultaneously *en route*, described our company as composed largely of women and children under the care of a spiritual adviser, hearty and healthy in appearance, wearing peaceful and happy expressions, and on our way to form a community in the wilderness, where our own forms of religious belief might be practiced without hindrance.

Memory of Portland is a happy blending — friends, beautiful drives, parks luxuriant with blossoming trillium and dogwood. At evening our little drawing-room was yet again heaped with roses, while once more a hearty Amherst cheer gave genial speed to parting guests.

Southward from Portland, Shasta is unmistakable king of all the great brotherhood. Intermittent snowstorms swept across, white clouds clung airily to his crown. Sunset light turned the snowdrifts rosy pink, like Mont Blanc from Chamounix.

Darkness brought the last evening on board the "A 1," and our affection for this delightful ten days' home was "done into rhyme" by Chief, whose ready gift at occasional verse was afterward in frequent demand : —

Valedictory Lines to "A 1."

You have carried us many a mile, " A 1,"
From the rising, away to the setting sun ;
O'er mountain and plain have we sped along,
With mirthful story and joyous song.
A happy crowd, without one " scrap,"
Save that gotten up by the newspaper chap.
For you we've ploughed snow, and filled your tanks,
And made you the scene of schoolboy pranks —
And you've filled our tanks, from many a plate
Placed by Lizzie and Charlie and Alfred " the great."
. . . .
In fact you're an A 1 car throughout,
And you know what you have on board, no doubt —
Where beauty and science and finance meet,
With " gyroscuti " as yet incomplete,
To eclipse all things that get in the way,
And at last to knock out Sol's dying ray.
But the rhymester grows sad as the time draws near
For parting — but then we shall reappear
On ocean's wave, and there's less regret
As we think of the cruise of the Coronet.

CHAPTER IV

SAUSALITO

Then is all safe, the anchor's in the port.
SHAKESPEARE, *Titus Andronicus*, iv.

WHERE was the Coronet? How would she look after her second voyage around Cape Horn? Every member of the expedition felt as vital an interest in a first sight of the fair craft as even her owner himself.

Nothing was seen of her on the way from Oakland across the bay; but at the wharf in San Francisco we were met by her sailing-master, Captain Crosby, and Frank Thompson, a young man who had charge of the instruments on the voyage. Both were brown and beaming after the four months' trip. Their report showed the Coronet still living up to her reputation for speed and seaworthiness. No accident had marred her record, the apparatus came in perfect condition, and she lay at Sausalito, a half-hour's ferry trip from the city, among the craft of the San Francisco Yacht Club.

Though intended solely as a pleasure yacht, the Coronet was found to offer unexpected generosity in space for stowing securely any farther amount of scientific paraphernalia. The more delicate bits of mechanism brought overland were soon safely packed on board, additional necessities being bought in San Francisco to avoid transportation from New York.

Weeks might have been filled solely with plans of hospitable friends for entertaining the expedition, and many invitations were accepted between visits to scientific headquarters and the adjustment of unaccustomed but graciously received cargo. It was a busy time.

Built up from the water, clinging to a steep hillside and embowered in foliage and blossoming roses, Sausalito possesses singular charm. From the narrow village street along the bay, steps innumerable lead upward past roofs of houses, past another tier of dwellings, to merge themselves in a gravel walk, still steeply ascending. Overhung by luxuriant trees and flowering shrubs, the " El Monte " was finally reached.

Not yet in entire readiness for her guests, the Coronet allowed them to gather for a few days at that little inn, — a place so distinctly foreign and picturesque that a shock of surprise always accompanied the unexpected sound of spoken English. A beautiful prospect rewarded the climb. Yachts lay at anchor in the bay, six or eight trading vessels and the Coast Survey steamship MacArthur, while beyond, villages nestled at the bases of hills, at this season green to their summits.

A tropic richness of vegetation covered the whole region, like one well-remembered June at Glengarriff. In San Rafael and other villages near Sausalito verandas were hidden in roses, the "beauty of Glazenwood" especially conspicuous in buff blossoming with shell-pink edges. Live oaks and the green bay, eucalyptus and sequoia filled the landscape, with palms and evergreens. Roses climbed often over high trees, hanging delicate blossoms from the topmost branches, a tangle of riotous flowering. Driving over the fine roads, Mount Tamalpais is nobly conspicuous.

Gradually ship's stores were sent on board, instrument-packing completed, the protecting paint holy-stoned off the deck, and staterooms put in sailing order. That assigned to the Astronomer and his companion was charmingly upholstered, both walls and furniture in rose-colored velvet. What feminine heart would not expand with gratified decorative sense, at the thought of thus voyaging daintily over the blue Pacific? Not unhappily I contemplated my modest store of silver wherewith to adorn the dressing-table in port, and a luxury or two planned for certain corners.

But the Professor's decorative instincts, while even keener than those of his household, — often, indeed, bringing original suggestions to bear upon the home habitation, — always take secondary place whenever touching the confines of scientific pursuit. Several improvements, therefore, of a technical and not wholly æsthetic character had soon despoiled the pretty pink room.

Raising the brass bedstead allowed nine deep drawers beneath, most useful during all the long trip. Two bookcases were fastened on the walls, and a case of twelve small drawers for lenses and eyepieces, plates and mirrors. A little curved sofa was also elevated in station, that under it a long box of like shape might be inserted, — invaluable for gowns and dress-suits all summer. A tall but sufficiently inoffensive wardrobe was made fast beside the closet.

It was all very snug and comfortable, with ample space for everything needed during seven months ; but the stewards looked on with despairing eyes as more and yet more of the rose-colored velvet walls disappeared ; and suspended telescopes were ornaments novel to the Coronet. Against the few inches of uncovered wall the Astronomer's protesting associate humbly tacked one or two portraits of her ancestors, her descendant, and certain home scenes, and thought her troubles over.

But shortly before sailing, the scientific head of the expedition appeared on board with a large mahogany case in which ticked loudly a sidereal break-circuit chronometer, which he calmly proceeded to screw to the dressing-table top, last rallying-point for dainty belongings. Descendant of two generations of astronomers and companion of a third, however, submissive attitudes of mind were inborn, so I smilingly assented to it all, even promising to wind that chronometer should such service become requisite by stress of circumstance. Though no longer a bower, the room was a sort of scientific emporium, the precious lenses had each its little drawer, and everything was in comfortable readiness.

A magnificent storm came up just before sailing. A wild gale beat the bay into white-caps, and set all the yachts dancing. Communication with shore was for several hours cut off; and even when the bay subsided into quiet, the sea outside still heaved tumultuously.

Just after luncheon and dozens of good-bys, on the 25th April, the Coronet sailed off, amid dipping flags and booming cannon, our own pennants flying, our farewell salutes waking Sausalito echoes. Out through the Golden Gate, across the bar (showing as a distinct line between pure blue of ocean depths and greenish, muddy waters of the bay), and into the broad Pacific the

Coronet tossed, where rear-guards of the storm still played with breaking white-caps out to a far horizon.

Finally the Bonita, which had accompanied us for a few miles to convey back to the city our pilot, a guest or two, and a dozen hastily written notes of farewell, changed her course; there was a last glimpse of a fast-receding shore; the Farallones were passed, and the expedition was left to itself in a wide waste of waters, with the Coronet for our two weeks' cosmos.

"Then the sun sank, and all the ways grew dark."

CHAPTER V

FIFTEEN DAYS AT SEA

Joyfully to the breeze royal Odysseus spread his sail, and with his rudder skillfully he steered from where he sat. No sleep fell on his eyelids as he gazed upon the Pleiads, on Boötes, setting late, and on the Bear that men call too the Wain, which turns around one spot, watching Orion, and alone dips not in the ocean stream.

Odyssey, v. 270 (Palmer's Trans.)

THE blue Pacific undulated gently, fair and sparkling; the voyagers lay lazily in steamer-chairs, with the deck gleaming white, brasses scintillating in the sun, white sails rounded with the wind, and motion just airy enough to exhilarate. In the shrill yet not unmelodious whistle of brown "goonies"[1] during these soft, bright May days at sea, could be heard potentially songs of orioles and bluebirds in New England orchards. Life lay dreaming in sunshine.

No throbbing engine stirred the heart of the pretty craft with restless pain and hot discontent,

[1] Great brown albatross always soaring round the Coronet when there was any breeze, and only rarely flapping their wings; but usually resting on the water when we were becalmed, paddling duck-like at the stern, and unable to rise except with much exertion, at first getting under way by running on the smooth water with extended wings.

but "quivering in the joy of her wings" she spread them like a bird to skim waves she scorned to plough through, tossing them off in foam from her bow.

Already I had twice crossed the Pacific Ocean by steam, yet its magnificent immensity was almost unappreciated until this voyage in a sailing vessel. Distance, if not annihilated, is at least mastered by latter-day triumphs of steam; but an indescribable charm lies in leisurely traversing enormous ocean spaces, dependent wholly upon the wind's sweet will; and when breezes depart, lying idly upon a glassy sea with sails hanging limp, a friendly sun flooding the decks with warm radiance, and a sky of softest, deepest blue brooding close above, affords one of the conditions yet remaining in this swift century when time seems of no value, and may be defied with impunity. As a rule, the Coronet voyagers were good sailors. Chief, an experienced naval officer, enlivened each meal with new and thrilling stories, and one of his inventions was a boon to the company,— a chess-board of ribbons woven upon a cushion, with pins in the bottom of the pieces to insure stability whatever the slope of the deck. Chief and Mrs. Captain, the Doctor and General soon became conspicuous experts, and many were the hours absorbed in this game.

Two or three days out a huge four-master loomed up superbly in the south, probably from Australia, sweeping on toward San Francisco.

Winds for a time were fitful, occasionally dying down to a flat calm. In lieu of anything more startling on these quiet days, the loss of a baseball overboard was brought into that category of noteworthy incidents. Twenty-three other balls below, provided against just such a catastrophe, were not enough to prevent an order to lower the dinghy, obeyed as promptly and with as perfect discipline as if the call had been "man overboard." Two officers speedily rescued the tossing white speck, the one lone object on the wide Pacific. But it had first to pass a careful scrutiny and much unsatisfactory pecking on the part of several inquisitive goonies.

A taste of brisker motion in the prevailing quiet fascinated another passenger to embark in the little boat, which then rowed off to a suitable distance for photographing the beautiful yacht. The sensation was unique enough for the risk of a genuine peril,—the whole Pacific Ocean with its broad and glassy rollers, a sense of immensity unparalleled, and the tiny dinghy, hardly an incident on its surface, our sole means of possible connection with the rest of the world.

These odd goonies were endlessly entertaining. Hundreds, even thousands of miles they flew over the waste of waters. Always voracious, they were easily tricked by trailing cork and fishhook baited with a bit of meat. The hook merely caught in their strong, curved bills, and they were hauled over the rail entirely unhurt, though always surprised at their sudden change of environment. Much flapping and screaming accompanied this operation, but once their web feet were set upon the deck, the birds were too heavy and awkward to fly back over the low rail; so they reeled about helplessly, or squatted flat on the white boards, occasionally spreading their wings, which fold curiously in angular sections. Weighing usually six or seven pounds, these albatross measured more than seven feet from tip to tip. When approached they objected audibly, snapping their bills with a sharp click. A purple and white ribbon was tied around the neck of one, which may yet be roaming the wide Pacific decorated with Amherst's colors.

Goonies were not our only visitors. One morning a tiny octopus, an unwilling caller, was washed on deck by a heavy sea and stranded. His head was surrounded by tentacles ending in suckers, — eight legs and two long feelers. He had a sort of bill like a parrot's, hard and sharp, and large weird eyes; perhaps a miniature edition of a famous character in "The Toilers of the

Sea." Sometimes, too, the brilliant flying-fish found themselves unexpectedly landed on deck,— bird-like and beautiful creatures, whose misfortune I deplored. A tiny Portuguese man-of-war was washed on board one day, a fairy bark less than an inch long, and full of shifting tints of blue.

For several days, in a region between the tumultuous winds off California and the steady trades farther south, light breeze or calm prevailed, tempting our men to a plunge overboard for an ante-breakfast swim. But a shark seen from time to time caused this exhilaration to degenerate into sunrise bucket-baths on deck, primitive shower baths with sailors for mechanism.

On Sundays everybody appeared in fresh white duck, and service was read in the cabin, a number of the crew being always in attendance, and adding their lusty voices in the tunes. "Eternal father, strong to save," that magnificent hymn for the sea, was a favorite feature.

One fair, sweet Sunday,

"So cool, so calm, so bright,"

a veritable

"bridall of the " sea " and skie,"

a breeze crept gently over the water, sails swelled hopefully — trade winds had begun. The great squaresail was set; stronger and more steady grew the wind to a full twelve-knot breeze, and for several days the Coronet fairly hissed through the water. The yacht deck is so much nearer the waves than that of a steamer that her speed, especially in darkness, seemed prodigious, as phosphorescent foam flew alongside, and a luminous wake trailed astern. Great following seas chased us, sometimes breaking lightly over the beam, but in the main slipping harmlessly beneath; the graceful craft, without a word of protest, sliding up to the crest, to float down hill again like a white seabird.

Except in rough weather, expedition work went constantly forward. As one delicate piece of mechanism after another was completed, all were brought for safe keeping to the once pink stateroom, and hung or nailed or triced up in every available spot. If one of its occupants chanced to throw out her hand carelessly in the abandon of dreams in the middle of the night, it was no uncommon occurrence to hit some perfected bit of apparatus and so set it off, to spin accurately through all the movements of picture-taking on its own account, or of evolutions which the half aroused sleeper dared not interrupt. Truly, science acquaints us with strange bedfellows.

The saloon was daily the scene of unwonted

activity. Doldrums had been passed, actually and metaphorically. Chief and Mrs. Captain spent sunny hours in fabricating small holders for endless plate chains; the Mechanician covered the big table below with mysterious devices in copper and steel, and the Musician experimented with different sorts of photographic baths. Such work as could be done on deck was always carried there; and by the time for afternoon tea, always served above, the entire party was generally ready to assemble on rugs and cushions in shadow of the mainsail, for an hour's listening before dinner to some entertaining book. The Coronet's library, full and carefully selected, had been increased for this voyage by friends and publishers until every taste might suit itself.

Exercise, too, was not neglected, and with more than eighty feet of clear deck, the number of laps necessary to complete the pedestrian's mile were often accomplished, and all sorts of hand over hand feats on taut halyards were performed, to the edification of the less athletic.

The picturesque habit of singing shanties [1] while hoisting the mainsail is still preserved among sailors on the Pacific. Finding that this

---

[1] The word, coming undoubtedly from the French *chanter*, has been perverted by unknown evolution to its present use and form.

EXPEDITION WORK ON BOARD

ancient though fast dying custom was thoroughly appreciated, our sailors gave many specimens, the mate singing a first-line solo, joined by the rest in a chorus following. With an accompaniment of such rhythm the big sail steadily ascended. An exceedingly interesting custom, with the peculiar hitch in the average sailor's voice, it is a performance not to be forgotten.

A number of these melodies became familiar, but the words were apt to vary with the soloist's ability to adapt current events to necessary metre. Versions in honor of the Coronet unfailingly brought a full audience. One of the most popular, with several sets of words, ran : —

1. Yankee ship comes down the river, blow, boys, blow.
  The Yankee ship comes down the river, blow, boys, bully boys, blow!
2. How d' ye know she 's a Yankee liner? Blow, boys, blow, etc.
3. Stars and stripes, and spangled banner.
4. What d' ye think of the Captain of her?
5. John L. Sullivan, Boston slugger.
6. What d' ye think of the chief mate of her?
7. Charlie Mitchell, English bluffer.
8. What d' ye think they had for dinner?
9. Monkey's heart, and donkey's liver.

10 Do yer know she's a Havre packet?
11 How d'yer know she's a Havre packet?
12 When she fires a gun, you hear the racket.

"Blow the man down" was also a great favorite: —

1 Oh, we are the sailors to join the Black Ball, uwa, wa, blow the man down,
   Oh, we are the sailors to join the Black Ball,
   Give us some time to blow the man down.
2 When Black Ball sailor get clear of the land,—

he has a variety of experiences emphatic rather than elegant.

With temperatures constantly warmer came evenings on deck, sometimes with informal lectures on astronomy illustrated by constellations conveniently at hand — or again quartettes sung by the light of swinging lanterns.

One of the company, whose energy needed some vent, planned a small paper, called the "Coronet Saturday Evening News," to which the reluctant company contributed articles, grave or gay, current or historic, — its society column especially brilliant, — and poems of much grace.

Though without cheerful submission to this draft upon intellectual resources, reading of the first number was greeted with much applause — from the contributors. In default of a press on board, transcription of this interesting sheet devolved upon the editor, who spent her entire day in the operation. Volume I. number 1, therefore, comprises the whole edition of this unique publication.

After the advent of the trades, daily runs averaged high: one triumphant noon record was two hundred and fifty-three miles; and night after night was full of the creak of woodwork and straining sails as the great boom tugged at the main sheets, and an occasional sea swirled along decks when the bow dipped into some watery mountain.

Toward the end of the second week, society around the yacht was increased by the advent of beautiful white birds, nautically named sea-hawks. Mother Carey's chickens, too, arrived, and marlinespikes with their two long tail-feathers. Even the goonies adopted fuller dress, now appearing with white bands around neck and tail. Over a brilliant blue and restless ocean, covered with flashing whitecaps, the Coronet was rapidly nearing Honolulu.

One big sailor developed a remarkable gift at telling astonishing tales without a shadow of foundation. Various members of our party often went forward to experience the enlivening influence of his talent for relating hypothetical incidents truly marvelous. Usually Big Jim's yarns were re-spun upon the quarter-deck.

Colossal drawn-work upon canvas in a variety of patterns was made a sort of leisure-hour occupation for the sailors, and afterward used in port to ornament the starboard gangway.

Another beautiful Sunday morning dawned, and with it a dim suggestion of cliffs and mountains on the far horizon. Off the port bow this faint shadow grew more distinct, until the barren slopes of Molokai came clearly into view, cut by enormous clefts, and streaked with tumbling cascades. Soon after Oahu, on which Honolulu is situated, rose above the waves, its rough, volcanic mountains sharply abrupt, and a little later dashing surf was discerned. As the Signal Station on Diamond Head came into view, we ran up four flags, K D J B, meaning "Coronet, New York."

A gorgeous sunset was flooding the world as this bit of official introduction took place. Great cumulus thunder heads were edged with dazzling gold; from a rift above, sun rays streamed over the rough peaks of Oahu and the uneasy sea like a huge inverted halo. Gradually the whole sky grew yellower, until everything was bathed in liquid gold; then the clouds broke into shreds, and the glory of the Lord came down and brooded over the waters.

Lights in Honolulu flashed out with darkness, one by one, and after the wide wastes of over two thousand miles of the lonely Pacific, it was friendly and homelike to know of other human beings near by, even on a remote cluster of ocean islands. Blue fire was burned for a pilot, who speedily responded with a tug, whose whistle that quiet Sunday evening announced our arrival to the city, already some days on the lookout for the Coronet.

Soon in warm, semi-tropical darkness, we were anchored in the narrow bay, with nothing of Honolulu apparent except twinkling lights and a dim mountain background, sharply serrated against the starlit sky.

CHAPTER VI

LIFE IN HONOLULU

The poetry of earth is never dead.
KEATS.

"PASS the first shower and turn to the right" — so runs the answer to inquiring strangers, desirous of reaching any given point in Honolulu.

But the rain seems to have a curiously unwetting character, like the swift downpours in sunny Bermuda; or else it possesses some attractive quality sufficient to counteract any unpleasant moisture. Nature behaves as if uncertain whether she is shining or showering, both rider and pedestrian sharing her indecision.

A fascinating city is Honolulu, embowered in tropical foliage fairly smothered in riotous vines, chasing one another in reckless race of crimson and golden and purple blossoming to the very tops of trees and buildings. Solid masses of color dangle high in air, and groups of Japanese and Chinese give a certain oriental effect to its thoroughfares. Native Hawaiians, the women in the prevailing white *holoku*, or unadorned "Mother Hubbard," throng the streets, and with some admixture of foreign blood are often handsome.

Architecture is simple and inoffensive, dwellings retreating behind wide verandas, so shaded by verdure that their modest lines are quite hidden. Portuguese houses may always be recognized by their attendant goat, grapevine, and tiny, naked baby. The rocks, chiefly volcanic, are too porous for building material, most of the native woods are too hard, and though a few edifices of a sort of coral conglomerate may be seen among occasional grass huts of natives, speaking generally houses are brought as timber from Oregon or Washington.

After mid-ocean coolness, the heat was noticeable, and at breakfast cream and fresh fruits appeared; while artistic water-jars, red "monkeys" of various shapes, adorned the sideboard. The awning was made fast over the quarter-deck, and staterooms were put into port order; even the obtrusive chronometer was taken on shore for rating, though telescopes still continued to adorn the once rose-colored room.

Paradise indeed, — the bits of coral and volcanic loveliness are rightly named. Lapped by gentle surf from the blue Pacific, fanned by trade-winds which steal away its fierceness from southern sunshine, singularly free from dampness, the islands are bathed in an ineffable glow of dreamy terrestrial atmosphere no less than in a certain poetic aroma left from the old, half-barbaric yet charming life of long ago.

Although the wonderful cloaks and helmets of yellow feathers once worn by royalty are now seen only in museums, there is even yet a suggestive national picturesqueness. Men lounging about wharves and corners wear hat-bands and decorations of peacock feathers, and chains (*leis* in the native language) of brilliant flowers about their necks. At a moment's notice any chance group can take up guitars or the little *ukulele*, playing and singing together in delightful harmony the half-plaintive and wholly sweet Hawaiian airs, with soft words like running water. A limp language, chiefly liquids and vowels, it is peculiarly suited to music. When the linguistic brook flows over a sharp pebble, usually a "k," it is for an instant broken into pretty ripples and flashes of sound, but it soon glides onward again, smooth and unruffled.

Society is distinctly American — constantly more so. American money is current, schools are founded upon our system, text-books published in the United States are used, and instruction is almost wholly in English. Recognized as the vernacular in 1876, it became compulsory in the schools; but even in earlier days it was studied by all high-class Hawaiians.

Uneasy political elements abounded. Royalists still hoped hopelessly for "restoration" and a limited monarchy, with its accompaniment of pleasant and characteristic court life; others looked for the return, as sovereign, of the popular princess Kaiulani, then in Europe, her cabinet composed of the best American element — a sort of amiable compromise. But as a rule the influential inhabitants earnestly desired annexation to the United States as a practical solution of vexed questions agitating the little republic. Since that hope long deferred is now, happily, an accomplished fact, their joy can almost be felt, meeting our cordial hand-clasp across the leagues of land and sea.

President Dole's charming manner, tactful administration, and personal popularity had apparently laid for a time the restless ghosts of political disquiet, but it was a period of waiting only. Effervescence seethed below the surface lull, and island politics were too complicated for easy disentanglement. To all our band of voyagers it seemed incredible that the powers at Washington should delay annexation of the fair islands, in general so ardently wishing it, so American in their development, with their wealth in sugar, in coffee, and in fruits, their persuasive climate, their endless possibilities. Almost from the moment of landing it seemed that the stars and stripes must soon float unchallenged from Government House. That and all the other public offices and residences were duly pointed out; all very much resembling sub-tropical edifices elsewhere in the world, but it hardly seemed to matter what anything was for, or who lived under any especial roof, when all over the city was such a rush of bloom and verdure, a commingling of delicious odors and flickering sun and shade from overarching palms and banyans.

Picnic making in Honolulu is a fine art. Open-air entertaining is constant. A lawn tea one evening at Waikiki, a suburb of the city, is still a sort of fairy memory. A low, verandaed house, far back among sheltering trees and vines, showed welcoming faces to the arriving guests, who were conducted to a sort of outdoor drawing-room (*lanai*), open on three sides to an enchanting garden close to the sea. Lapping gently against the white beach, summer ripples almost reached the algarobas in the sand, whose feathery foliage threw delicate shadows from the western sun. On the grass, light tables stood about, each with a bowl of plumeria or other characteristic flower; a larger table at one side was covered with bright *leis*, fragrant coffee and dainty refreshment. One of our hostesses had been an old friend in Washington, years before, when her husband was Hawaiian minister to the United

States. Strolling or sitting in groups under the trees, with *leis* (bright flowers for the ladies, a sort of green laurel for the men) thrown over their shoulders, the guests were served by softly stepping Japanese maids.

Toward the city tall cocoanut palms stood out on a point of land in silhouette against the yellow sunset. On the other side rose Diamond Head, bleak, barren, impressive in the purpling east. As twilight crept onward, Japanese lanterns began to gleam here and there among the shrubbery; but no trace of chill or dampness touched the air, and on the darkening sea horizon the southern cross burned in white splendor.

A quintette of native musicians sat in shadow, playing the *ukulele*, a banjo, two guitars, and a *taropatch*, occasionally singing Hawaiian melodies full of surface gayety and lightsome rhythm, yet soon revealing an unsuspected undercurrent of deepest pathos. To the gliding music two or three friends, for our edification, danced native figures on the grass — strange and graceful. All danced for a time in the drawing-room, but the dim lawn, the sweet, haunting music, and the lapping waves cast an unresisted spell, and the company soon drifted out under the algarobas.

Late in the summer night a happy party drove back to the city, and were rowed out to the Coronet at her anchorage in the dusky harbor.

Valleys are numberless, an especial charm of the islands. There picnics most abound. One fair, fresh day we rode on horseback through the city and up Manoa Valley; leaving the horses at a Chinaman's domain, a short walk through banana groves brought us to a rushing stream, whose banks, thickly covered with rich vegetation, rose steeply against the blue sky, secluding the little camping-ground. Distant hillsides were exquisite with bluish-green atmosphere and caressing sunshine.

Picnics in other environment than valleys were no less lovely — on famous beaches where we first had surf-baths in a rainbow-tinted sea, afterward sitting upon the grass for luncheon on closely woven native mats, the making of which is almost a lost art. Picnics were given on mountain-tops, upon verandas and in gardens — at Laiakanoe hale (Point of Mists) near Pearl Harbor, where the whole American navy may now make itself at home, with the Waianea mountains (Watchtowers of the West) forever upon guard. The sweet, simple, gracious life of the islands is delicious even in retrospect.

Surf-riding is an exciting amusement; native boatmen, each with a happy passenger in his canoe, paddle out beyond the breaking waves, only to ride beachward on a rushing, foamy crest. Bathing-suits are necessary for the passenger (the islander does not trouble himself with one) because overturns are not infrequent. If the foreigner can swim until the native comes to his aid, all is well. Most persons can do so, and are generally rescued alive, though not invariably. Still, the perilous pastime continues in unabated popularity.

Ladies use no side saddles. With full, divided skirts the Hawaiian method of riding is not less graceful than our more accustomed fashion, besides being more hygienic for both horse and rider. Tennis and golf clubs add their testimony to a certain fresh tonic in the balmy air. Yet the purely native character is undeniably indolent, amiable, and careless of the morrow, with an untroubled satisfaction in the sunshine and bananas of this life. To the average American manner its southern softness and grace are added, producing a charm too frequently absent from more prosaic conditions at home.

The President and Mrs. Dole were unwearied in personal kindness to the expedition. A breakfast, among other pleasant attentions, was given us at the executive mansion one dewy morning, amid palms and birds and flowers. The dining-room was wide open to veranda and garden, full of summer scents and murmurs, heavy shade of bread-fruit trees, and sound of dripping fountains. The first six courses were fruits, — alligator pears, *papaia*, fresh figs with cream, mangoes, pomegranates, and more familiar strawberries, bananas, and oranges. Another morning Mrs. Dole invited to her beautiful home thirty or forty friends, members of an informal literary club, to meet the guest from over-seas. With none of the harassed ferment and eager attitude characteristic of that objectionable type, the genuine new woman, these native-born ladies of American descent were an audience appreciative to an inspiring degree. Discriminating in their criticism, they showed the gracious effect of careful study in conditions of untroubled leisure.

Sojourners in this enchanted land are always taken to the *pali* (precipice). Back from the city climbs the road, through Nuuanu valley, between curious peaks and ridges of volcanic hills, ten or twelve hundred feet high, and past roadsides abounding in bright lantana. Scarlet and orange and yellow, it is always at first greatly admired by visitors, conspicuous in their buttonhole bouquets of the gay blossoms. But no resident would be guilty of disporting himself in the flowers of this overrunning pest, supplanting as it does worthier growths, and causing great wrath in the innocent breasts of husbandmen and householders. It is, however, not less decorative because objectionable to agriculture.

Brightly green in afternoon sunshine rose the

RESIDENCE OF PRESIDENT DOLE IN HONOLULU

valley's inclosing walls, their summits shrouded in soft cloud, often condensing suddenly in swiftly passing showers. Carefully cultivated fields of *kalo* (or *taro*) showed each root of handsome leaves set off by itself in a little hill surrounded by water. Personally I could wish this highly useful plant might be kept exclusively for decorative effect, wherein it is a success; since as basis of *poi*, the national food, it becomes an unappetizing edible of barbarous qualities.

In 1795 the Napoleon of Hawaii, Kamehameha the First, fought a great battle near the present Nuuanu road in his final conquest of Oahu, one of the last islands to acknowledge his supremacy. His enemies fought bravely until their leader Kaiana was killed, which utterly discouraged and soon laid them low; while the remnants, forced up the narrowing valley before the victor, were finally driven over the *pali* at its head, 800 feet into the plain below.

Looking backward for an instant from this battlefield, the city lay bathed in warm sunlight; far beyond the blue sea, hazy with distance, gleamed to a shimmering horizon. The valley closes in yet more narrowly as the road continues to ascend, and at last a low wall ahead apparently bars farther progress, and giant sentinel towers of rock rise several hundred feet on either side. No premonition of approaching grandeur touches one's expectation ; only some pretty vista is anticipated, like scores of others the world over. But beyond that wall the scene might well be in some novel planet, so rare and radiant, so shining and peaceful, so far and grand — its effect was too overpowering for more than the first exclamation of delight.

Directly below the parapet falls a steep precipice. At its foot is a serene and sunny country bathed in unspeakable peace after æons of unforgotten volcanic agony, — stretching indefinite miles to right and left, and joining northward the pale and misty sea, with white surf breaking high on many a rocky point, or creeping silently up to silvery beaches curving around distant bays. Over all, brooding sunshine, pensive in still beauty; close at our left an amazing pinnacle of reddish volcanic rock, hundreds of feet above. Curving sharply to the right, and descending steeply under a perpendicular wall, the road zigzagged downward to sea-level.

No words even suggest the strange grandeur, the foreignness, the exquisite beauty, the illimitable pathos of this *pali*. Its charm "vanishes in the writing, and remains dumb in the telling." But in my innermost heart of memory it dwells for all time. During an instant of joyful awe it seemed that this world lay solemnly in the very presence of God.

With return to every-day emotions once more, consciousness of the furious wind grew unpleasantly insistent; and a native boy, carrying a violin and riding a much decorated horse, passed by and down the steep path, with never a glance at the outspread glories, but many an interested one at the strangers.

A unique sight in Honolulu is the magnificent Bernice Pauahi Bishop Museum, which contains the finest collection of South Sea Island specimens in the world, an epitome of Polynesian ethnology and natural history. Founded in 1889 by the Hon. Charles R. Bishop in memory of his gifted wife, herself a direct descendant of the Kamehameha line and actually heir to the throne, the nucleus of the collection was Mrs. Bishop's own store of mats, calabashes, and distinctively Hawaiian relics, bequeathed to her as sole survivor of the original royal line, and supplemented by bequest of Queen Emma's treasures. Later collections made in New Guinea and New Zealand came to its shelves, and now the whole story of Polynesia may be read within these remarkable walls.

The Kamehameha schools for boys and for girls, established by the will of Mrs. Bishop, are still farther monuments to the extraordinary generosity and wisdom of this unusual woman, and to her husband's well-directed liberality. Native girls in airy, comfortable recitation rooms are carefully taught subjects of probable use in after life. They sang for us American songs, occasionally one of their own quiet melodies with soft Hawaiian words.

Far back in the misty annals of this little group of famous islands, women here and there emerge from a gray past in bright relief,— welcome incidents in a monotonous story of conquest and rebellion, war and victory. Even in prehistoric times wives of chiefs played conspicuous parts; and in half fabulous tales of old voyages, the hero-chieftain took in the great canoe his wife and his astronomer, — evidences of good taste and sagacity in that twilight period of Pacific island history.

Astrology was practiced, and its devotees continually studied the heavens, the places of moon and planets in relation to especial stars and constellations being deeply associated with the fortunes of many high families. Navigation by the stars was constantly practiced. Not only at sea were women brave and helpful; but warriors' wives often followed in the rear of armies, carrying food and water, and sometimes aiding the belligerents more practically. Manono, wife of a brave and popular young chief at the head of an insurrection, fought by his side, continuing her part in the battle when he fell, finally herself dying upon his prostrate body.

Rank, too, descended through the mother, and marriages of high chieftains were carefully regulated. A queen's son was a noble, no matter of what class his father. On the other hand, the son of a chief, if his mother were a person of no especial rank, would be one of the masses like herself. For state purposes, therefore, great care was used in contracting marriages, and offers were frequently made by women. In 1807 three men were put to death because the head queen of Kamehameha the First (Keopuolani, recognized by all as the highest living chief) was dangerously ill. She, respecting the sacrifice, recovered and lived sixteen years, surviving by four years her illustrious lord.

In later days, too, women are prominent. As queen regent Kaahumanu was an enlightened ruler, a promoter of education and good morals. Living until 1832, her reign, if it may be so called, was full of progress and prosperity. Kinau, as premier in Kaahumanu's place, used her strong influence for law and order. In 1859, that brilliant king Kamehameha the Fourth and his charming wife Queen Emma founded the hospital bearing her name, which still keeps her in no less loving remembrance than if she had been elected chief ruler, as at one time was possible. When King Kalakaua died, his sister Liliuokalani became queen; thus once again a woman held the helm of state.

And so, onward through all the years from the brave wives of early chiefs, generations of Hawaiian women are incentive to every native girl of to-day. Always prominent in island affairs, they have now a better opportunity than women in many other nations to live up to their inherited traditions, and carry on a worthy island story.

The old native church, for which each stone is said to have been contributed by a different and devoted Hawaiian, is quaint and attractive upon the exterior, and its service is conducted in the soft syllables of the "boneless" language.

Life in Honolulu harbor had its own distinctive interest. Anchored far enough out to avoid the dust and heat of the wharves, there was always a gentle breeze under the awning of the Coronet's after deck, where all our meals on board were served. Flowers filled the saloon, fresh fruits were unlimited, and our time-bells and those of the U. S. S. Adams, as well as of merchant ships lying near by, mingled unanimous hours and half hours musically all over the harbor, as days and nights rolled on.

Naval officers are always charming hosts. Enclosed with flags, a native orchestra discoursing sweet and plaintive music for American dancing, flowers, summer gowns, cool refreshments,— the decks of the Adams were often the scene of gay teas and receptions, the Coronet's gig and the naval launches carrying festive parties from one to the other, and the shore.

Entertaining, too, in its way, was the artless family life in progress upon a neighboring big merchantman. The captain with his wife and three small children were very much at home upon their nautical abode; and while the ship was overhauled for repairs, hammers ringing out as the old paint was chipped off her huge sides, a fresh coat following closely with rejuvenating effect, father and mother played with the baby, or wheeled it up and down the deck in a small carriage, while two older children pirouetted about in little sunbonnets, — citizens already of the maritime world at large.

Much of the Astronomer's time was spent in rating the chronometers on shore, in observation of transits by night, and in farther tests of the new glycerine clock in the Surveyor General's office by day.

But in spite
>"Of hours that glide unfelt away
>Beneath the sky of May,"

and the delightful simplicity of life in Honolulu, we were not oblivious to the complex problems abounding in the island. Native customs are slowly but surely dying out, and an Americanized future is now inevitable, — more useful if less picturesque.

CHAPTER VII

HAWAIIAN VOLCANOES

> The reticent volcano keeps
>   His never-slumbering plan;
> Confided are his projects pink
>   To no precarious man.
> Admonished by his buckled lips
>   Let every babbler be;
> The only secret people keep
>   Is Immortality.
>
> EMILY DICKINSON.

REMOTENESS of the Hawaiian islands from one another is hardly appreciated by those who have never visited our new possession. Honolulu and the island group are synonymous to most persons. Usual maps, too, give suggestion that channels at most separate the islands, which may lie an hour or two's sail apart. Local steamers, however, require two days and a night for the trip from Honolulu southward to ports on Hawaii, the largest island.

On the twenty-first of April, Mauna Loa, 13,700 feet high, had begun to show lurid red above its topmost crater (Mokuaweoweo), betokening one of its infrequent eruptions. From a hundred miles away at sea enormous pillars of red flame could be seen streaming upward. During nearly three weeks a magnificent spectacle had continued, and the island papers were filled with details of the new activity.

But the projects pink of this particular volcano, no less than of its generic brotherhood, were concealed from every precarious man; and no one dared to foretell Mauna Loa's never-slumbering plan. Its reticence was complete. Though the fires were evidently growing less, such an opportunity must not be missed. Honolulu fascinations were hard to leave; still, the inter-island steamer Hall at its next departure had several members of the expedition on board, as well as a number of friends from the city who joined us for the week's trip. The Coronet would repose peacefully at anchor during this side excursion, with those on board whom urgent expedition business aided in resisting the volcano.

A few residents of the city, and a German, Dr. Friedländer,[1] had already made the ascent; but such hardships are encountered that few persons attempt it. Mr. Dodge of the Government Survey had been one of a party to reach the summit, and his description was a truly thrilling tale. He told us that the cold was intense, ice filling gaps and chasms over which they climbed, a heavy

[1] "Kilauea," by Dr. Benedict Friedländer, *Himmel und Erde*, vol. viii. p. 105, December, 1895.

snowstorm was in progress, and mountain-sickness attacked many of the adventurers. Horses, too, suffered greatly,— one dying in the rough upward scramble over masses of *pahoehoe* and sharp *aa* (lava).

The Wilkes scientific expedition round the world in 1844 had made the ascent, and their trail, still dimly defined, had been found by Mr. Dodge and his party at about 11,000 feet elevation. They remained over night upon the edge of the crater, whose walls vary from 500 to 700 feet in height, while the lake of liquid fire was not less than 1600 feet in length with a width perhaps two thirds as great.

From this appalling expanse two huge fountains of flame a few furlongs apart were seen to spout upward thousands of tons of lava, brilliantly lighting the whole crater, and the heavens above. Their average height was about 250 feet; but frequently spurts or fiery jets would fling red-hot bombs to a much greater elevation, while the boom and roar of this whole inconceivable outburst filled every pause in the wild wind. Smaller columns constantly leaped forth in different spots, occasional whirlwinds carrying pillars of smoke hundreds of feet above the walls, and lifting great slabs of hardening lava only to cast them off again. The edge of creeping lava, brilliantly red, lapped a margin of white snow.

Spray from these upspringing fountains of flame made graceful curves as they descended in sparkling showers, while at their bases a crimson sea seethed and boiled like angry surf upon the shores of Hades. Descriptions of the activity of these fountains perhaps suggested in small measure the terrific happenings at the surface of the sun.

What climb could be too arduous for a view of such scenes!

As the Hall left her Honolulu moorings, the wharf presented a characteristic sight. Native men in picturesque hats trimmed with bands of peacock feathers, women in the universal holoku, boys in no particular costume to speak of, and everybody draped in wreaths of flowers, filled all available space with an amiable crowd. Little two-wheeled vehicles waited in the background, full of pretty children and young girls in white; close by, handsome brown boys dived in the clear green water for dimes and quarters thrown from the steamer. Flowers were everywhere, tropical sunshine and good-humored faces. Slowly receding from the wharf, the Hall passed the Coronet at her anchorage, acknowledged her parting salute, and turned south toward the incomparable volcano.[1]

---

[1] "To no other volcano can Mauna Loa be compared in its vast mass, or the magnitude of its eruptive activity." — Captain C. E. DUTTON, U. S. Army.

Slight roughness in currents of the inter-island ocean caused the passengers, regardless of nationality, to subside unanimously. The southern shores of Molokai are more nearly level than its rugged northern coast. Lanai was passed. But no interest in topographic features sufficed to stir the occupants of the forward deck, Chinese and Japanese, Portuguese and Kanakas, in every imaginable half-breed combination, — all lying with their luggage around them, in picturesque confusion. Small Japanese babies with shaved heads and fringe of hair and Chinese infants with tiny queues diversified the scene, but made no sound. So thickly was the deck covered with various reposing nationalities, all in their native attitudes, that stepping room was out of the question. It was a motley array. The captain was a handsome, swarthy islander, the stewards light-footed Japanese.

The level light of sunset turned the whole great slope of Maui brilliant red; deep shadows were thrown into enormous gorges; occasional patches of brilliant green sugar-cane appeared, the tropical effect emphasized by tall cocoanut palms near the shore. Moist, filmy clouds hung about the mountain peaks, now and then drifting off aimlessly. Many natives, draped in *leis*, were leaving the Hall at the little town of Lahaina, and through the purser's politeness we took a closer view of an unfamiliar hamlet.

Crowds filled the landing-place, sandy streets were bordered by banyan and cocoanut trees, and a pond hid itself beneath the crimson flowers of some greenly spreading water-plant. Boys walked calmly up the straight, columnar trunks of trees, bringing back cocoanuts as spoil; women and children played in the sand. Along the beach lay *waa*, curious native canoes with extended outriggers; and surf beat high on lava reefs outside.

The evening was warm, the breeze soft, and her deck a charmed spot as the Hall steamed away in early twilight.

Kailua, a place of much historic interest, was passed at dawn, too early for landing. Miss Field was reported as still pursuing there her studies into the condition of natives.

About noon of the second day, Kealakeakua Bay was approached, and the monument at the village, Kaawaloa, in honor of Captain Cook, surrounded by a fence of chains and ancient cannon. The discovery of these islands by the famous navigator was the turning-point in their history, ushering in a new era of prosperity. He first visited them in 1778, his second visit being in the autumn of the same year. In January of 1779 he anchored in this bay, where nearly a century and a quarter later an eclipse expedition tarried on its way to the Orient.

Cliffs nearly five hundred feet high rise straight up from the sea, and around the bay, with water as clear and green as an emerald, nestles the little town. The site of an astronomical observatory established by Captain Cook near by was not seen, but it was a thought full of interest that instruments had been so early set up and observations made in this far-away harbor of Hawaii.

The murder of this sturdy explorer, 14th February, 1779, so affected the world at large that no foreign vessels attempted to anchor there for over seven years. Land for the monument, erected in the name of his countrymen, by Lord Byron, commander of the frigate Blonde, was given by the Princess Likelike (Mrs. Cleghorn, sister of Queen Liliuokalani).

Telephone service is nearly perfect, and one may speak from any of these little native towns to all others on the same island. Everywhere we eagerly asked for news from Mauna Loa. Each reply was more discouraging than the last, — its fires were no longer visible; but ever hopeful we voyaged onward.

All along the Hawaiian shore lie occasional villages; here and there a freshly made cave in the cliffs showed a late burial-place of some native. In earlier years a popular custom, this method of interment is now infrequently practiced. Innumerable natural caves indent the

HAWAIIAN VILLAGE LANDING-PLACE.

rocky coast, against which deeply blue water beats itself into tremendous walls of white, the spray flying high and startling the air with a resounding boom.

This southwestern coast of Hawaii is barren but impressive. Apparently one huge lava flow, it is only in spots overgrown by verdure. Black death and green life lie side by side. These oases hold a few grass huts, and half a dozen straggling cocoanuts, while far above lies the great mountain, its top lost in drifting cloud. With a base so enormous, and slope so gradual that its true proportions are difficult to conceive, Mauna Loa gives almost the impression of being flat on top. Around Hawaii the sea is very deep, and if its mountains were referred to their true bases at ocean's bottom, where the range really rises, they would be no less than thirty thousand feet, or nearly six miles in vertical height.

The national vegetable is *poi*. I had tasted this article prepared in several ways, always with a new sensation but without marked satisfaction. Noticing upon the Hall an old friend and resident of Honolulu, engaged in drinking from a glass something evidently cold and possibly good, I boldly ordered a similar beverage, only to find it gray and elastic, and alarmingly acid in flavor.

I had eaten black bread in Germany with nameless accessories, raw fish in Japan, unclassified crustaceans, and shoots of bamboo, and national dishes in various other regions of the earth's surface. But liquid *poi* is a discouraging delicacy which outranks them all; and a single draught required all the cosmopolitanism I could summon to refrain from unseemly demonstration. The scenery again resumed its greater charm, with one fleeting glimpse of Mauna Kea, high in the sky.

Toward evening the Hall anchored off Punaluu, two hundred and seventy miles from Honolulu, where landing appeared an uncertain performance, through pounding surf, still encircling the coast in walls of swaying whiteness. About nine o'clock two small boats shot through from the little town, to bring our company ashore. They were propelled by natives ready to dash back with us into lines of breakers at the most favorable instant.

Great rollers chased us madly, raised the boats like egg-shells in a wild rush forward, then broke over the lava reefs with resounding din on either side, now and then enveloping us in heavy showers. Still the native oarsmen kept peacefully onward, guiding their craft with much skill through the narrow passage among rocks, taking each sea just right until both boats were brought up safely beside a little wharf in partial shelter of a small bay, where one assisted jump landed the voyagers among a variegated crowd watching our arrival with interest. It was called an unusually calm landing for that port.

Punaluu society seemed to be in a state of primitive simplicity. After walking up to the little hotel through a path between lily-ponds bordered by rushes, no proprietor was apparent. Everything was open, all on one floor, doors and windows hospitably wide, beds carefully made, and not a soul in sight. So we took possession, and after a time an excellent Chinaman appeared who officially turned the house over to the guests and their peaceful slumbers.

CHAPTER VIII

A HAWAIIAN JOURNEY

According to her cloth she cut her coat.
DRYDEN, *The Cock and the Fox.*

AH HE provided a delicious breakfast; during intervals of dreams we had heard him chasing fowls of different species far into the night, and the merry company remained unsubdued despite heavy clouds enshrouding Mauna Loa, and occasional sprinkles. But impending disappointment as to the great eruption became sad certainty. The fierce fires had wholly withdrawn into deepest mountain recesses; not a tint of red remained to suggest the unconquerable energy which for nearly three weeks had flashed forth in sublimity. The climb, therefore, to Mokuaweoweo, nearly fourteen thousand feet in air, was reluctantly abandoned, since the grand spectacle had seethed itself into rest, and nothing except a dark crater would reward the ascent, views of distant islands and ocean being almost constantly cut off by clouds and mist. Kilauea, less than one third as high, but always interesting, and evidently preparing for an eruption of its own, became our volcanic Mecca.

Something, called by courtesy a train, inaugurated the first stage of the trip. The track, supposedly about two feet wide, varied pleasantly in breadth, sometimes more, sometimes less, each rail wandering goalward at its own sweet will. An amusing toy engine, old and rusty, with a shrill shriek to correspond, drew one small platform on wheels where the satchels of the company reposed under a canvas cover; a second, with benches for the men, and a third, boasting an awning under which the ladies were invited to ride in magnificence.

Across a volcanic country staggered this establishment, in amazing curves, brushing past cattle grazing on scanty grass among the lava. Everywhere superb white poppies and a scarlet flower like salvia bloomed lavishly. Over deep gullies, in the rainy season grotesque lava beds for rushing streams, around hills, skirting miniature valleys, the little railroad pursued its uncertain way to Pahala Plantation.

Here the sugar-making process was watched — from cutting the cane on four thousand acres of land, to the completed crystals, not sufficiently refined to prevent free importation at San Francisco.

Horses and a rickety stage were soon in readiness. Several ladies rode in divided skirts, after the sensible island fashion.

A desolate country, indeed, this leeward side of Hawaii, covered with lava hardened into weird shapes like nearly stagnant waves, too lazy to flow, but which, just as they were curving for another lap, stiffened into crawling circles, or heaped in chaotic masses. Brilliant yellow and scarlet milkweed blossomed along the way, standing decoratively against black lava backgrounds, while armies of brown butterflies which might have recently escaped from some New England meadow, hovered near, perfectly at home in these foreign uplands.

Pele, goddess of fire, seems appropriately fond of red, for red flowers, red leaves, red berries, and red birds abound on the slope toward her citadel. But she is a jealous divinity, and no flower or berry must be picked on the slopes of Kilauea, for that would imply admiration of them. And all homage must first be paid herself — else she will send rain or other damper to pilgrim enthusiasm.

A native woman, Kapiolani, having decided in favor of Christianity, journeyed to Kilauea in 1824, purposely to set Pele at defiance. Upon the very edge of the crater this brave woman, trembling at heart, we must suppose, if only from hereditary dread, performed various acts designed to excite wrath in the fiery goddess, whose power for centuries had been thought absolute. Yet no fatality followed these impious performances, the truly heroic attempt justifying itself. But superstition is hard to dislodge, and to this day few natives would willingly pluck anything on the way to the crater. Pele's particular flower, the *ohia*, grows on large trees, a magnificent fringe of scarlet like flame, apparently the long, clustered stamens of a tiny, whitish corolla.

After the charmingly hospitable manner of islanders, we were entertained at luncheon at Kapapala Ranch, a garden of beauty midst great barrenness; and here another vivid account of the great eruption was given. From the depths of a heavy snowstorm a benumbed and half-fainting company had watched through the night the gigantic spectacle, listening to the never ceasing roar of flames and internal seething of this indescribable cauldron. Though that was but a few days before, when ships at sea could view the pillar of fire on Mauna Loa's crest for one hundred and fifty miles, now it was all out and gone — not a breath more of this terrific energy, and only a calm summit reposing peacefully above, innocently laying its huge crest against the sky, even in its great height like any New England hill.

After luncheon riders and stage traversed the short grass, still upward, a telephone wire the sole suggestion of direction, or an occasional intermittent line of lava wall. In the native language going toward the mountains is called *mauka;* to the sea, or anywhere away from the heights, *makai.* Sometimes a few lichens half concealed the rock they decorated; or a vine full of blue morning glories climbed high on rugged masses, and a few ferns grew here and there. The brilliant milkweed went to seed at will, — showing, in friendly association with buds and blossoms, a silvery puff ready to float on the first inviting breeze. Running about among the rough masses were tiny quail, and a large rock had a round hole in its side, through which a fern seemed to have pushed itself, turning skyward, and unrolling its little green knapsack in fitful sunshine. Behind, the blue sea tumbled in white surf on a jet-black beach of volcanic sand; before, the lava-covered heights we toiled to reach.

After the Half-way House, kept by an odd character with amusing conversational powers, a few more ragged and scrawny trees appeared, but this side of the island is undeniably desolate.

The great lava flow of 1868 came down the southwestern slope; we had passed it, now a black and cold devastation, in the Hall at the little town of Kahuku. That eruption, beginning in the summit crater, was accompanied by all sorts of convulsions. Earthquakes shook houses quite off their foundations, breaking vases and china indiscriminately; so that now dining-room closets in this uneasy region are fitted with little guards in front of every shelf — like a ship's galley. A "mud-flow" swept downward a league's distance in a few minutes, not less than half a mile wide and thirty feet deep. So rapid was its engulfing rush that thirty or forty persons were overwhelmed, and hundreds of animals perished. A tidal wave, too, toward fifty feet high, rolled against the coast, killing nearly a hundred persons, destroying villages, and permanently submerging the road at Punaluu, where landing is now effected.

Another great lava stream broke forth late in 1880, and flowed down the eastern slope for many months. Sometimes a mile and a half in width, it slowly but steadily approached the town of Hilo, causing much depression in dwellers there, and in the price of real estate. The speed of these streams of *pahoehoe* is so moderate that time is sufficient to remove livestock and portable treasures from its path. But one may not transport his sugar plantation, nor his house and gardens; and property continued to depreciate, as this slow, deliberate, relentless stream came nearer and nearer. In the general panic lands of ancestral memory were abandoned for a song. When within three quarters of a mile of the town, and destruction seemed only a matter of days, then it was that trust in ancient superstitions once more prevailed, and the Princess Ruth, a member of the old royal Kamehameha family, went out with a company of friends to appease if possible the wrath of Pele. Bottles of brandy and gin, pigs, chickens, silk handkerchiefs, and locks of hair were thrown into the sluggish stream with appropriate ceremonies.

It must have been a weird scene, as described to me by an eye-witness, a participator in the evening's events; and the party returned to town, confident in the success of their diplomatic mission. Singularly enough the flow stopped next day, dividing itself and dying out harmlessly. But the real estate could not be bought back by its former owners. Neither gratitude, nor terror of Pele, sufficed for that.

Late in 1886 hundreds of earthquake shocks were felt, and soon after New Year's an eruption occurred at the summit crater of Mauna Loa, accompanied by a lava stream following in general the great flow of 1868. The death of Princess Likelike, sister of King Kalakaua and Liliuokalani, occurring about the time this flow ceased, gave abundant reason to many natives for its ending, a certain propitiatory offering thus implied. In December, 1892, Mokuaweoweo was again brilliantly active for three days; but the frequent minor eruptions are subordinate to the two disasters most frequently quoted — in 1868 and 1881.

As the old stage rattled on, the advance riders quite out of sight, a damp mistiness hung over the landscape. Pele's scarlet flowers glowed in the dull day like veritable bits of her own flame, and native stories, legends, and myths beguiled the rough and ever upward way. Curious blowholes abound, where bubbles of lava once burst, leaving deep caves and pits. In one of these tradition says Pele once came in search of a pleasant home. She was accompanied by sixteen *hula* (dancing) girls, who thoughtlessly picked the delicate fringed blossoms of the *ohia*, and otherwise showed carelessness of her peculiarities; so that returning one day to the cave after an excursion about the mountain, she suddenly turned them all into pillars of *pahoehoe*, while a central one of *aa* is commonly reported to represent herself, the only bit of that formation in this vicinity. Even now no native can be persuaded to enter that cave without an offering, if only a *lei* for Pele.

Stories relating to fissures, caves, or lavablocks were told on the way, unfailingly picturesque, and showing a strong bent of the native mind. "Henry Gandell's Leap" is a wide crack on the slope of Mauna Loa, which a man riding in hot haste down the mountain-side saw too late to avoid. Spurring ahead, he took the flying leap, and landed safely on the farther side; but the strain caused his ultimate death a year later, though the horse was reported as still alive. Belief in *kahunas* (witch doctors) has by no means died out; and a prophet named Lukula foretold that a corpse (*kupapa-u*) would arrive from over sea, bringing death and desolation; that a great eruption would then occur, to be followed by a royal restoration. Cholera came and Mauna Loa broke forth, but restoration remains unverified. Toward evening a cloud of white vapor indicated the nearness of Kilauea; and steam bursting out of holes and cracks in the ground all about, very hot and fringed with exquisite sulphur crystals, formed the suggestive approach to Volcano House. Perched high on one wall of the enormous crater, this hotel commands the entire scene of desolate grandeur. A lake of black lava, three and one half miles in one direction by one and one half in the other, is surrounded by nearly perpendicular walls varying from seven hundred and fifty to three hundred feet high, at one end of which a secondary crater sends out volumes of sulphurous fumes. Mauna Loa rises grandly toward the west, and in the north Mauna Kea raises a rugged peak yet higher.

Fortunately without other guests the Volcano

House hospitably offered ample quarters, most attractive of all a long billiard-room, dark raftered overhead, with huge stone fireplace decorated by lurid scenes of volcanic activity. A friendly house-party took immediate possession, and many and startling were the stories told round the blazing fire; for evening air at high altitudes is cool, and in Pele's very sanctuary any tale is credible.

CHAPTER IX

KILAUEA

After eruptions are over,
 After the mountain is dumb,
After the fire has vanished,
 Up to the crater we come;
Wander on black lakes of lava,
 See the white steam rising higher,
Gaze at the calm crest of Mauna —
 After the fire.

EARLY dawn revealed a weird scene. Steam issued all about the house from countless fissures. Below, the enormous lava lake stretched grim and motionless, partly hidden by mist. Over it fell showers at intervals, while the hotel stood in brilliant sunrise light, and a superb rainbow arched the black pit from wall to wall with heavenly radiance. Then sunlight broke through the last shred of mist, chasing away the shadows, and it was sweet, sparkling morning on Kilauea.

Descent into the crater, on the sure-footed horses, is a memorable experience. Back and forth the narrow path winds, down the wall, through masses of ferns and foliage, until the great cliff behind shuts out half the sky, and high in its thick verdure a single bird-song came out airily into the dewy morning. But the lake was unspeakable desolation,— black lava in writhing, curling, creeping masses as far as the eye could reach, growing hotter to the feet as the steaming secondary crater was approached, until a stick thrust slightly into a crack came out in flames. Strange contradiction of this fast-dying nineteenth century, a telephone wire crosses this Hades of desolation, and, though useless now, once served to connect the Volcano House with a little building close to the active crater. But in the picturesque native language, "it was eaten up by fire;" that is, the hot lava walls caved and the little house fell in.

Every tinest crack sends out heat like a register when the furnace below is red hot — only in Kilauea one may not send down impatiently to know why the check drafts have been forgotten. The crater is a bewildering mass of tumbled lava, hissing sulphur steam, and unbearable heat. Another great eruption was daily expected.

The various kinds of "blow-holes" were curiously fascinating, with heat too intense even to stand near in many cases, yet tempting as far as endurable. Anxious as usual for new experiences, I descended a short distance into one about fifteen feet deep, but speedily returned, nearly overcome by sulphur fumes and a temperature of unimaginable degrees Fahrenheit.

"Yes," said the General sadly, "she could n't stay as long as she hoped, but her next article will be entitled 'My Interview with Satan, or What I saw in the Blow-hole.' It will be very exciting and sufficiently authentic."

Such amenities materially mitigate the dreary grimness of a slumbering but restless volcano, and the active crater's extreme edge was cheerfully approached. The soles of our boots were already too hot for comfort, and prevailing sulphur odor was variegated by a strong smell of burning woolen, as folds of a gown rested for a second against an unsuspected crack in the flaky and shining black surface. Thick fumes concealed the pit activities six hundred feet below, and a slight change in the wind would have brought suffocation in its train. Above the whole surface, even of cooler portions of the lake just traversed, the heated air lay in a quivering mass, and retreat was a gradual return to comfortable breathing. As the lava became cooler, away from the more active pit, a tiny green fern now and then peeped hopefully from an unaccustomed bed; and enormous encircling walls, often nearly perpendicular, rose above with grand but sullen effectiveness. The necessarily slow climb to the upper world made a mile or two of free wild galloping on the mountain horses a subsequent necessity. Sulphur fumes were blown off in brisk

SULPHUR BLOW-HOLE IN THE CRATER OF KILAUEA

breezes, a handful of wild roses was gathered, and a soft-footed Celestial announced luncheon just as we drew rein under the tree ferns by the hotel entrance.

Kilauea-iki is well worth the short walk through unfamilar fields, past strange holes going down in blackness to unknown depths, but fringed on the edge by luxuriant ferns. For a mile or more a footpath winds pleasantly through characteristic vegetation, emerging suddenly at the brink of a huge pit a mile across, sunk over seven hundred feet into the earth. The sides of this enormous bowl, in places very steep, are covered with shrubs and low trees, and far down were wild goats peacefully browsing on the margin of another black lake, dead, cold, its waves stiffened in immemorial ripples. This weird spot lies quiet in the unmoved calm of centuries, no eruption having broken its repose within historic time. Four or five holes in the bank, however, are said to have afforded exit for small streams of lava which as lately as 1844 sped downward in molten cascades, — travesty upon murmuring brooks hastening to join some still, green, forest tarn.

It is a strange region, full of uncanny interest; but afternoon tea on the Volcano House veranda brought familiar modern life once more to the front.

A stroll to gleaming sulphur caves beyond the

hotel gardens gave renewed consciousness of the proximity of nature's vast, uncouth forces, often slumbering but never inoperant, and constantly ready to burst forth into sublime activity. The sparkling yellow sulphur crystals are exquisitely fragile, and the cracks they fringe, emitting steam and smoke from fires perilously close at hand, are too hot for thorough examination. Trees and shrubs near these vents are incrusted with yellowish deposit, making spectral pictures.

The Olaa road to Hilo traverses another world. This little town is about thirty miles from the Volcano House. The government road thither is hard, perfectly kept, and bordered by tropically luxuriant vegetation. Banks of tall tree ferns, shrubs showing both white blossoms and scarlet berries, huge creepers (*ieie*) with long, lance-shaped leaves, hanging their blossoms high in air from trees by which they climb, composed a tangle lush and bewildering. The water-lemon with decorative leaves, blossoms like a passion flower, and oval, purple fruit; bananas, so sheltered that their great leaves are perfect — not whipped into ribbons like those on windy Bermuda; a large bush with drooping, greenish white bells, called *floriponda*, made the whole roadside a joy to traverse, and without the haunting pain that everything might be barbarously cut down before the next visit.

Careless guardians of highways in the United States might well take to heart the practical lesson from Hawaii, where intelligent as well as æsthetic appreciation protects and enhances every roadside beauty. The district (*aina*) about Hilo is now used extensively in cultivating coffee; and although the land costs only five or six dollars an acre, five times that sum is used in clearing it for plantations. But no settler is allowed to bring his fields to the roadside; a border of natural growth must be left, not less than one hundred and fifty feet wide, between his coffee plants and the highway. He may keep four hundred feet (along the road) of open lawn before his house. Otherwise no border growths must be cut, and I hear the prescribed width is now increased by one hundred feet.

With natural loveliness so jealously guarded on every hand, I could but remember certain drives in New England, once fair and beautiful, now reduced to scenes of painful devastation; where farmers, in temporary lack of occupation, might have been seen cheerfully mowing banks of maidenhair, chopping down trees, blackberry and elder bushes, pulling up clematis by the roots, and setting fire to the pathetic remnants. Here in remote Hawaii every tree and shrub, flower and vine, is watched and cared for; and a total dearth of advertising signs on rocks and conspicuous places is enforced by law.

Dwellings on the Hilo road are simple to primitiveness, but with unfailingly attractive grounds, often laid out by those consummately artistic landscape gardeners, the Japanese. Garden walls covered with a happy riot of nasturtiums; walks and steps of tree-fern trunks, brown and elastic; wide verandas, even if the house might boast but a single room, furnish the foreground; while luxuriant forests, laced in a tangle of lusty vines, approach close behind the little ranches.

One tiny house not over ten feet square was nearly smothered in rankly gorgeous vegetation: roses blooming lavishly, tall begonias in full flower, hedges of callas, tree ferns, *floriponda*, coffee plants, wild roses, bananas, *ohia*, actual trees of coleus — all thriving in very wantonness of summer life, hid roof and veranda in clinging embraces. Manifestly belonging to a coffee plantation, a sign on the gate announced that "the owner does not wish to show this coffee, but any gentleman desirous to learn, and not actuated by motives of curiosity, can see it by application to" some one within.

All the ranches were named, in the soft, invertebrate native words. Letter boxes stand at gates, wide open for incoming or outgoing mail. Nothing is lost.

At a little inn halfway to Hilo, where luncheon was served in the open air, the entire party were greatly exercised, not to say profoundly shocked, by a small child of the Portuguese couple in charge. This promising youth, eldest of three, was just two years old, in a white cambric dress and big sunbonnet, with innocent blue eyes and flaxen hair; yet that depraved infant was an experienced smoker! Holding between his rosy lips the stump of a cigar abandoned by some guest, he sauntered past the newest comers, puffing vigorously at what remained, occasionally with two dimpled baby fingers removing it from his lips with the air of a smoker of fifty, while he blew uncertain spirals into the sunny air. The child's father seemed to think it an unusual accomplishment — in which opinion we were gladly unanimous; but he could not be made to understand its danger, only smiling foolishly at remonstrance. The pale little Portuguese mother hardly appreciated the Doctor's energetic remarks, though she finally caused the cigar to be taken away, whereat the child wept dismally, with rising anger, and refused reconciliation.

Primeval tropic forests crept up to the inn on three sides, and the mynah bird was ubiquitous. A large bird with conspicuous white spots on its wings, imported years ago from Jamaica, in hope of eliminating certain destructive ground-worms, it has at last filled the islands. Flitting decoratively about among the green was a little scarlet bird called *elepaio* from its song; and the native *omao*, greenish in hue, sang a full, liquid note.

On this side of the island it rains so much that two hundred and thirty inches often fall in a year, so the owner of a coffee plantation assured us. A greater contrast certainly could hardly be imagined than this "mad extravagance and splendid luxury of nature," and the barren coast skirted by the Hall in reaching Punaluu — a landscape of lava-flows.

With tropical sunshine by day, it was nevertheless the big fireplace which attracted us at evening. The last before returning to Honolulu brought music, a little dancing, extracts read from guest-books full of odd and interesting entries, and a story or two, told as the logs burnt into glowing red coals, and stars came forth in sudden clearness from a misty sky.

Later, the last vestige of fog disappeared, Mauna Kea's rugged peak rose in the distance, and grand Mauna Loa came forth unshrouded. Behind his majestic shoulder Jupiter was setting close to a crescent moon, and almost unearthly stillness lay over the world. Far below the crater was smoking vigorously, and close at hand the ground at every pore breathed white steam, quickly absorbed into a dry and silent night. Grass and ferns were full of insects singing or chirping or scraping their nocturnal music — little songs in the grass which, emphasizing the silence, might have seemed in far-away Massachusetts, but for the surrounding scene, so foreign to that placid land. Yet even here, on distant Hawaii,

> "A minor nation celebrates
> Its unobtrusive mass,"

and the same sky overarches all.

CHAPTER X

A POI LUNCHEON

<blockquote>
Those palates who . . . .
Must have inventions to delight the taste.
Pericles, i. 4.
</blockquote>

As rice is the national dish of Japan, so a certain vegetable concoction already mentioned, called *poi*, has that high distinction in Hawaii. Prepared in a variety of ways, each, to the average visitor, is less alluring than the other.

When the members of a native family are seen happily clustered about a large central bowl, dipping contentedly therefrom with two fingers (under some circumstances three) a viscous substance of nondescript color, which seems largely composed of an indifferent quality of yeast and mucilage, one may be tolerably sure they are indulging in the questionable delights of this delicacy.

A charming invitation to partake of a *poi* luncheon, given in our honor, and under most delightful circumstances, had been accepted at a ranch on the way back to Punaluu. Perhaps the unfavorable verdict on a nation's staple might be reversed.

In a radiantly sunny morning good-bys were said to Kilauea's height, and the old stage, saddle horses and riders, and two adventurous pedestrians started downward, past dewy wild roses, accompanied by countless bird-songs under a sapphire sky. So distinct was the crater floor, seven hundred feet below, that its very lava wrinkles could be seen; the sulphur cracks steamed incessantly close at hand, but the slopes of Mauna Loa lay clear and unveiled in early sunshine, without so much as a bit of hanging cirrus on the crown to suggest a lingering suspicion that his great fires might still be ready to spring forth in renewed splendor. The mighty mountain grew more impressive with every hour; and a belt of cloud halfway up the peak added to the apparent height of Mauna Kea.

Vegetation became more scarce as Volcano House was left farther behind on the road to Kapapala Ranch. Pele's scarlet flowers on their scrubby trees glowed finely in the morning brightness, and young shoots low on the ground, called by the natives *liko lehua*, showed all their topmost leaves in no less brilliant masses. Small blue flowers known as *ioi* grew on tall stalks all the rough way, and songs and stories varied the jolting ride.

An oasis amid volcanic desolation, the Ranch seemed a charmed spot, even lovelier than at our first visit. Within its outer gate bright green grass and a few old trees greeted eyes weary of endless acres of *pahoehoe;* still farther, the house nestled in gardens like some tropic flower. Fuchsias hung their blossoms high above our heads, avenues of tall coleus led into mazy labyrinths of bloom, and friendly welcomes awaited us on shady verandas.

For the benefit of guests unaccustomed to Hawaiian ceremonies, everything was done in a style as distinctively native as might be consistent with comfort. In an *imu* (underground oven) beyond the garden, young pigs and chickens had been cooking for two or three hours, delicately wrapped in *ki* (or *ti*) leaves against red-hot stones, between layers of vegetables — the whole covered with earth. Men, experts in an art now dying out, were removing from the pit the various edibles so daintily cooked that they hardly held in shape while transported to the house.

Luncheon was served on the wide *lanai* (veranda), each chair being thickly draped with *leis*. Roses were everywhere, an undesirable insect which has nearly exterminated Honolulu roses not then having reached Hawaii. To be thoroughly native, the company should have had no chairs, but it was not deemed necessary to submit the guests to so thoroughly un-American a position as sitting flat upon the floor; so that single detail was omitted.

The luncheon was lavish, even without reference to *poi*, served in various forms. The proper method of eating it with two fingers from a bowl was successfully imitated; still, though more palatable than before, one of the guests continued to regard this vegetable with suspicion. Its color is against it, granite gray not being an attractive tint in articles of diet.

Under the inspiration of the feast many picturesque tales of life in the old days were told. Hours for state calls were from three o'clock in the morning until nine or ten, and royalty wore superb capes and helmets of yellow feathers. Each bird (the *o-o*) had but two tiny tufts of these feathers under its black wings; and as this decoration began to extend gradually to persons of lower rank than chiefs, the plumage soon became very scarce. When the bird was caught and his golden ornaments pulled out, he was set free, without that slaughter of innocents practiced in more civilized lands. As time went on, flowers for personal decoration came into general use, thus probably originating the graceful custom now shared by all classes of wearing green or flowery *leis*.

Young girls of high family attended missionary schools, being taught many useful arts and precepts; one maiden was especially impressed by three rules of conduct: in after life she must neither dance nor drink wine, nor must she do anything without her husband's permission.

When this little Hawaiian girl, married soon after leaving school, arrived in Honolulu as the bride of a chieftain, the Queen waited upon her at dewy dawn (while the bridegroom paid his respects to the King), inviting her to breakfast at the palace — a gracious royal command. But true to training, she replied that while it would give her the greatest pleasure to accept, she could not do so without first asking her husband — a form of answer entirely novel in all the royal experience.

When healths were drunk at state banquets, the little bride still remembered her instructions, and refrained from touching her glass, a surprising performance to the king, at whose side she sat. But when the young husband finally learned of these eccentricities, he speedily reduced the three rules to one, no less definite. Hastening at the next banquet to obey him when a health was toasted, she innocently drank the whole contents of her glass at once. The remainder of that feast now lies in her mind as but a confused shadow of memory.

A charming little boy in whose veins runs the blood of many nationalities had listened all his life to tales of past days told him by an old chief; he repeated many of them after luncheon in his sweet, childish voice, the following one written as he narrated it, in his own words: —

"The son of King Kamehameha shot a great many arrows at the bread-fruit trees, which took away their juice and spoiled them. So he had to be sent away to the island of Lanai, and there he found some hobgoblins who planned to kill him. They asked him where he was going to sleep that night, and he said, 'In the big waves.' So in the night they went out to the big waves, but could not find him, and were drowned.

"In the morning the others asked him why he did not sleep, as he had said, in the big waves; and he replied they were so large he decided to sleep in the little waves. The next night they asked him and he said, 'In the big thorns.' So they went after him, and a good many were stuck on the big thorns and killed. And when the survivors asked him in the morning why he was not there, he said, 'The little thorns were more comfortable.' Then he decided they must be trying to kill him; so the next night he got the rest of the hobgoblins into a house, and they thought it must be for some grand entertainment; and then he stuck all their eyes together with breadfruit and burned up the house.

"So after that he was safe."

As the pleasant shadows lengthened, kindly friends gathered under the trees at the gate, sending cheery *aloha* (farewells) far down the grassy road, as the uncertain stage bore us once more to the outer barren. Surrounded by friends and flowers, a deep-blue tropical sea, vast volcanic mountain slopes, and the soft, sweet atmosphere of enchanted Hawaii, even *poi* became poetic in retrospect.

Again we were in sight of the blossoming white poppies of Pahala, the rusty little engine having waited until we chose to arrive. The sea stretched blue to the horizon, white surf still tumbled grandly on the black beach; and after Ah He's appetizing supper, another exciting trip through the breakers (again, a so-called "quiet" sea, which merely did not wholly capsize the boats) brought us on board the Hall, peacefully anchored beyond the rush and roar of waters and encircling reefs. All night our steady way was ploughed northward, past the barren Hawaiian shores toward our first landing the next afternoon.

Near the beach at Kailua lie the ruins of a lava fort, built by Kuakini long ago; and great Kamehameha the First once lived here in a grass hut, on the site of which Kuakini (called by foreigners Governor Adams) built later a house of lava and cement, the broken walls of which are still standing. Once a large native population filled this town, and a missionary church, whose square tower rises near at hand, is the first built on the island. A large house with double verandas is still the property of the Queen Dowager Kapiolani, widow of Kalakaua.

But in addition to historic remains, as we sipped cups of tea in the shade of an airy lanai, we witnessed a unique sight — the apparently cruel native method of bringing half-tamed cattle on board for shipment to Honolulu. Confined in small pens or yards on the beach, one or two at a time are first lassoed; and with men on horseback in front dragging them with main strength by a rope attached to their horns, others behind cracking long whips, the terrified creatures are driven, galloping madly, into the surf, and forced to swim out to small boats waiting beyond the breakers. Tied to the edge by their horns, still in the water, usually eight on each gunwale, they are rowed slowly out to the steamer, and hoisted on board by block and tackle. Half-drowned and quite subdued by fear and pain, they stand in long, shivering lines, on the lower deck; if a horn breaks off or pulls out, no matter. They will be killed in a few days.

Native houses and straggling vegetation, with great Hualalai rising over eight thousand feet in the background, afforded characteristic setting for the lively scenes on the beach. Natives in big, picturesque hats wreathed with flowers were riding recklessly back and forth on high saddles, a variety of animals were adding their own voices to a composite chorus, — squeaking, crowing, neighing, bellowing, squealing, — and children covered the sand; it was a gala day. Palms stood up tall and tropical in the warm air, and soft-lying cloud began to drift low down on the mountain-side.

Great lava-flows and barren shores on the homeward trip looked more familiar since we had penetrated the very heart of a country strange with sharp contrasts and endless charm. And now we were leaving it — to Hawaii's weird island good-by; with its grimness, its sublimity, its steaming promises and fiery fulfillments, its tropic beauty and black devastation, a long good-by.

CHAPTER XI

WITH KATE FIELD

Death is the crown of life.
YOUNG.

A fiery soul, which, working out its way,
Fretted the pigmy body to decay
And o'er informed the tenement of clay.
DRYDEN.

AT Kaawaloa more cocoanut-palms and natives; and a small boat put off from shore, bringing Miss Field on board, weary with arduous research into the condition of the native islanders. Lack of proper food and attention, a severe cold contracted through exposure to varying temperatures at different altitudes, and general fatigue had left obvious traces on her pale face.

"Riding too hard," the purser said, after he had shown her to the stateroom she had reserved.

Miss Field's wide acquaintance, the interest in her felt in all parts of the world, and constant questions as to her last hours on earth have caused the hope that as I was with her during that memorable time, although an experience of deep pathos amid prevailing light-heartedness, its narration may not prove inharmonious, but welcome, even if sadly so, to many hearts.

Comfortably settled in her berth, Miss Field asked that our good Doctor, whose merciful service was in constant demand for ills more or less serious, should come in to advise about her health. Very serious after his few moments' chat, he reported that she would enjoy seeing a caller. Having had but slight acquaintance with her, I nevertheless accepted her invitation, being warmly, even enthusiastically greeted. Extreme pallor had given way to bright but feverish color. To an unprofessional eye she looked better.

"Oh," she exclaimed, "it is such comfort to be on a boat again, though I usually hate a boat; but to be going somewhere actually again, and to see white people once more, and up-to-date white people at that! I have seen natives, natives, until I am completely worn out!" — her naturally brilliant manner beginning to reassert itself.

"Talk about the quiet and pleasures of the country," she went on. "It's the noisiest place on earth — chickens cackling, roosters crowing, dogs barking at all hours!"

The natives themselves and the political situation she discussed warmly.

"Too much education of the masses," she asserted. "The public school system is responsi-

ble for a great deal of evil, just as it is in America."

"On the frequently quoted principle," said her companion, "that it spoils a great many good cooks, and makes a superfluity of poor teachers?"

"Exactly," she answered. "It's all a mistake. But they are lovely, amiable people. I've enjoyed Hawaii, but I am pursued by Kamehameha's fishponds. I can't strike any settlement on the island but that one of those malarial holes is pointed out to me.

"When I was at Kailua I did think they would be intelligent enough to avoid them, but I had no sooner arrived than I began to smell malaria, so I knew there was another historic fishpond close by."

She had evidently talked as much as she ought, but as I rose to go she remonstrated.

"It's such a comfort to see you," she said, pressing my hand. "I am only tired all out. Riding all sorts of horses (for my own got a sore back), and tramping over their lava beds and looking into the condition of these natives. Riding astride is all right, but there can be too much of it. Yes, I am too tired to do any more just now."

She lay back with her cheeks very pink and began to ask about our expedition to Japan, in which she seemed greatly interested.

In passing Keauhou, where Kamehameha the Third was born, a handsome native came on board to see Miss Field. There is no white family in the town, and the Hawaiians there called her *Kela wahine naanao* (that learned woman). He did not remain long on board, and she was persuaded to rest quietly for a while. Toward evening I made another short call, during which her characteristically sparkling way of putting things was unusually manifest. As it grew dark, a few native Hawaiians gathered on deck near her door, singing sweet and plaintive melodies, accompanied by guitar and *ukulele*. I asked if it disturbed her.

"Oh, no," she answered. "Music is Paradise to me, and I shall sleep all the better for it."

And, indeed, she did sleep through the evening, apparently with much peacefulness; but the Doctor, seeing that she grew worse, stayed all night by her side. About two o'clock a decided change occurred, and early in the beautiful morning he told me that he had been fighting for her life ever since she came on board, obstinate pneumonia his antagonist. With little hope, from the first, of conquering, he had continued to give her stimulants on the chance of sustaining the slight strength remaining. He thought she must have had the disease for several days, while still exposed to constant hard riding and all temperatures. Naturally it had made irrevocable headway.

The truth was very hard for me to tell her — that in all human probability she must die before another sunset. Miss Field listened in almost a dazed way at first. Then she said, —

"Yes, yes — give me time. I must think of so many things." She lay back for a moment in strange stupor, while I quietly waited. At last, arousing her gently, —

"Miss Field, you would better tell me the names and addresses of any friends to whom you would like to have me write," I said, wishing fervently to aid in some strong way the energetic soul still struggling to keep manifold interests within a loosening grasp.

"Yes, yes, I must," she replied, giving me an address in Washington, which she spelled out carefully and accurately. Then she began to dictate a letter, clearly enough at first, but soon confused.

"It will need a lot of editing," she finally said wearily, while fragmentary sentences relative to her work for the Chicago "Times-Herald" fell at intervals. The Doctor continued stimulants, but she sank more and more deeply into unconsciousness.

All through the morning she aroused a little as I spoke, but it was evident that she was rapidly dying, and her breathing became very labored. As we passed Maui she suddenly opened her eyes and looked out. The cliffs are bold and rugged, and the mountains very impressive, with cloud-shadows chasing over them, and between island and steamer lay a bright blue strip of white-capped sea.

"Oh, how beautiful!" she exclaimed, and for a moment her eyes brightened clearly.

Holding her hot hand, and longing unspeakably to give her a little human love to reach heaven on, I sat there all the sunny, sparkling morning. A few necessary addresses and bits of practical information were plucked at intervals out of the rising tide of death's lethargy, when suddenly Miss Field looked up with entire naturalness.

"What did you say was the name of your expedition, and what are you going for?"

"The Amherst eclipse expedition," I replied, "and we go to Japan to observe a total eclipse of the sun August 9th."

"The Amherst eclipse expedition," she said brightly; and those were her last words on earth.

She simply slept more and more soundly as her soul drifted farther out on unknown waters. All this time the captain of the Hall had been pushing the steamer to the utmost, to reach

COTTAGE IN DR. McGREW'S GROUNDS WHERE MISS FIELD DIED

Honolulu if possible before Miss Field should die.

As we neared the harbor all her scattered belongings were put together, — saddle, whip, walking-shoes all scratched and scarred with rough lava — even her possessions looked tired and helpless, lonely and discouraged. We landed much ahead of usual time.

Soon after the Hall came alongside the wharf, a stretcher was brought from the Adams, upon which Miss Field was tenderly carried to the residence of Dr. McGrew, a friend who had been very kind during her entire stay in the city. In a beautiful open cottage under the palms in his grounds, she peacefully stopped breathing a few minutes later — a sad home coming for us to our fair Coronet.

Next day a large and appreciative company gathered in the Central Union Church, to say good-by to the earthly presence of this bright woman who had yielded her life pathetically in behalf of a strange people. As the casket, heaped with tropical flowers, was carried out, the organ softly played "Home, Sweet Home," and thought of her real home, after years of brave and unremitting effort, brought unaccustomed tears. Miss Field had never acknowledged herself defeated, and who shall call this unfinished work and lonely death defeat — in face of an illimitable future?

CHAPTER XII

A MID-PACIFIC COLLEGE

Tutor'd in the rudiments
Of many desperate studies.
SHAKESPEARE, *As You Like It*, v. 4.

WITH our last days in Honolulu, the fifty-fifth year of Oahu College was closing; for so early in the history of their peaceful conquest of the islands, begun in 1820, did the fathers of civilization think it necessary to broaden their educational resources.

Liliha, wife of Boki, the then ruler of Oahu, was evidently a woman of force, if also of energy misdirected. Plotting to overthrow and remodel everything in general, the government incidentally, she has left a somewhat unenviable record. But feeling on one occasion unexpectedly generous, she joined her husband in presenting to missionaries the site for Punahou school, now Oahu College. Barren and unproductive then, no one could have foreseen its present tropic beauty. Liliha's portrait represents her leaning affectionately upon the shoulder of her lord — he in mighty helmet, she in a necklace of human

BOKI, RULER OF OAHU IN 1820, AND LILIHA HIS WIFE

hair. Of the two, her face is decidedly the stronger.

From its modest beginning fifty-five years before, the institution has steadily grown in scope and influence. And now another building was to be dedicated to high ideals, — beautiful Pauahi Hall, yet one more gift of the Hon. C. R. Bishop, whose liberality seems limitless. The ceremonies were held in the new hall, on the evening of the 21st of May. The fine building of native stone, with its semi-tropical style of architecture, the brilliant electric lights, the polished hard-wood interior finish, and the paintings, etchings, casts, and books, and the band of musicians, were far from presenting what the average American would have imagined a typical scene of mid-Pacific civilization.

The Rev. Daniel Dole landed on the island, 21st of May, 1841, and began his work of instruction and enlightenment. A school was opened the same year, with a small class of children, in a little adobe building a few yards east of where Dole Hall, built in 1848, now stands; and this was the real beginning, the birthplace, of Punahou School and Oahu College, of which Mr. Dole was one of the founders and the first head. This ripe scholar and Christian gentleman, father of the late President of the Hawaiian Republic, gave an impetus and tone to the school which caused its pupils to take high rank in whatever college they might subsequently enter in the United States. Habits of accuracy and literary taste were as valuable then as now, and these were bestowed in liberal measure at Punahou. The name means "new spring," and this it became in all senses. The high thinking of those early days must have meant very plain living, for the pupils paid but fifty cents a week for their board.

A permanent schoolhouse was opened on the 11th of July, 1842, — a building of one story, the ground plan like the letter E, inclosing two square courts, with schoolroom in the centre. This building, also of adobe, its timbers and rafters of wood from lovely Manoa Valley, roof of thatch from Round Top, and plaster and whitewash from coral limestone and sand of Kewalo reefs, was purely a native product. An opening with about a dozen pupils between the ages of seven and twelve was not a very striking inauguration, but it marked the happy point when children would no longer have to be sent around the Horn for an American education, spending years away from lonely parents.

In 1854 the school became a college, not with rank corresponding to Yale or Amherst or Williams, but carrying the student about to the junior year of those institutions, and equipping him with peculiar fitness for the more liberal development which they could offer. In 1863 nearly one hundred and thirty acres of the land of Punahou, with buildings and improvements, were deeded to the trustees of Oahu College. Its most constant and generous patron has been Mr. Bishop, whose devotion to the interests of the institution as well as to every noble cause in the islands is a conspicuous factor in its history and success.

By 1864, when President Mills resigned, the college had been placed upon a self-supporting basis, though the genuine and happy turning-point in its fortunes occurred in 1881 at the celebration of its fortieth anniversary, when a large fund was raised by alumni and friends. In 1882 another large sum was added to the building fund, and the following year the main building was erected, in 1884 the Bishop Hall of Science, and in 1885 the new President's House. By 1889 the endowment fund received $56,000, of which about two thirds had been given by Mr. Bishop.

Shortly after Mr. Frank A. Hosmer, of the class of 1875 at Amherst College, became President, the semi-centennial was appropriately celebrated, in 1891. The orator of that occasion was the late and well-beloved General Armstrong, who in a brilliant and characteristic speech gave many incidents of old days when he was a Punahou boy; while Professor Alexander, the distinguished historian of the islands, and for seven years President of the college, told its story in his own delightful manner.

Since then Oahu College and its preparatory school have gone steadily forward, becoming more of a power with every year. In 1893 President Hosmer suggested, in view of increasing needs of the institution, that a new academic hall be built; plans and drawings were submitted by various architects, every design, however, carrying out the idea of a solid stone pier rising from the foundation to form a tower for a telescope, thus giving all possible stability to an elevated observatory. A compromise between two of the plans was made, the result being a very effective and handsome building costing a little less than $80,000, — another superb monument to Mr. Bishop.

An island of volcanic rock, varied by a few coral reefs around the edges, is not the most prolific spot for good building materials, most of the stone being porous and not impervious to water, while the beautiful *koa* wood is so hard that it is impracticable on account of expense of working it. Since timber for all frame houses is brought from the American coast, a comparatively simple house is of much greater cost than in this country. Many specimens of native stone were submitted for use in the new Pauahi Hall, and the building committee finally accepted a compact gray stone found at the entrance of Manoa Valley, not only very handsome in itself, but giving evidence of entire power to withstand water.

The grounds, with their mass of tropical foliage, the fine algaroba trees, and avenues of palms, were in gala dress for the dedication ceremonies, and the formal transference of Pauahi to the college faculty. The address of the evening was given by the distinguished President of the Republic, the Hon. Sanford B. Dole, who was greeted with prolonged and enthusiastic applause. His delightful speech was full of the best spirit of modern Hawaii, reaching always for the highest, yet permeated throughout by the poetry bequeathed from the older days.

After the address, the keys of the new building were delivered, with an interesting speech by the Hon. W. R. Castle, to President Hosmer, who responded with feeling tribute to those devoted men in the past who made possible the development of to-day, a growth probably far beyond what they would have dared to dream in the simple beginnings of their time. A fine organ, presented by Mrs. S. N. Castle as a memorial to her husband, was played during the evening, and there were selections by the College Glee Club and an orchestra lately inaugurated by the students.

The wonderful Hawaiian climate, never too hot and never too cold, appeared that evening at its best, and will always add its indefinable but no less haunting charm to Oahu memories.

Strolling across the grounds under the tropical foliage and by the light of swinging Japanese lanterns, we reached the President's House, where an informal reception was held.

This only night on shore at Honolulu was followed by a lovely morning, dewy and fragrant, amid trees and vines, flowers and shrubs of the college grounds, musical with bird-songs, and recalling the choicest of New England's midsummer dawns. A day full of meaning to Oahu College was this last one for the students in the old historic building, quite inadequate now, yet full of tender association. Their feelings were not of exultation merely, in entering a wider life in modern environment. Eager and intelligent faces, and the appreciative attention accorded a short talk given them at prayers, betokened a waiting future full of progress and achievement.

CHAPTER XIII

THE LEPERS OF MOLOKAI

> But Sir Launfal sees naught save the gruesome thing,
> The leper, lank as the rain-blanched bone
> That cowers beside him, a thing as lone
> And white as the ice-isles of Northern seas
> In the desolate horror of his disease.
> LOWELL, *The Vision of Sir Launfal.*

SIR LAUNFAL "gave the leper to eat and drink," and despite the poverty of his repast, it seemed to the gray and gruesome recipient like fine bread and red wine —

> " For a god goes with it, and makes it store
> To the soul that was starving in darkness before."

And notwithstanding their ignorance and uncleanness, the Hawaiian lepers are treated with care and generosity deserving more than grateful recognition from a glorified community. Indeed, it is said that life on Molokai is now considered so desirable by many natives that they have been known to feign the disease in order to be taken there, supported by the Government in ease and idleness.

Leprosy, not indigenous but imported, was first observed in the islands in 1853. When its spread, in 1865, was thought alarming, an act was passed isolating cases in separate establishments. A year later about one hundred and forty were sent to Molokai, but rules were not very strictly enforced. If one only of a married couple developed the disease, the other was allowed to go also to Molokai. On accession of King Lunalilo in 1873, strenuous efforts were made by his new cabinet to carry into effect a law of absolute seclusion, and over five hundred persons were sent to the settlement. This of course excited bitter opposition, but it was in line with the enlightened policy of this monarch, who lived to reign only a little over a year.

Now, although healthy wives or husbands may not accompany their diseased consorts to the settlement, marriages on Molokai among the lepers themselves are not forbidden. Some children born in that retreat are actually healthy, and without trace of the dread disease. When on their official visits the Board of Health bring back to Honolulu all such cases, if the parents consent, and they are reared and educated away from infection. Often they do not develop leprosy at all in after years. If the unmistakable signs appear, they must return to their birthplace. What a weird and terrible meeting between parents and children so tragically reunited!

When the Board of Health start for Molokai, heartrending scenes often occur as the steamer is about to sail, — friends and relatives of lepers crowd the gangway, begging permission to visit afflicted comrades. But quarantine is necessarily strict and unswerving.

Without seeing practical means of gratifying his desire, the Doctor had always hoped to visit the leper settlement. He remarked pathetically that any suggestion of his wish was far from popular on board the Coronet, being met either with stern silence, or browbeating and discouragement, — even by assurance that he would certainly be thrown overboard upon his return, should he finally succeed in reaching Molokai, goal of his hopes.

But in this often unreasonable world where Fortune brazenly chooses her favorites regardless of merit, sterling worth, probably by mistake, is sometimes rewarded. One of the customary tours of the Board was due a day or two before the Coronet set sail for Japan, the good Doctor received a cordial invitation to join the medical men on their trip, and regardless of a dire fate upon his return he accepted with alacrity.

Upon the unfortunate lepers Government spends annually $150,000, or one tenth of its entire income. The Doctor's own journal, which he has kindly given me, is only second in interest to a personal visit. He rowed away from the Coronet on the 22d May, . . . "after receiving all sorts of warnings and good counsels, and scrambled up on the wharf of the Inter-Island Steam Navigation Company. Already a few passengers had arrived, and some officers of the Board of Health were there to keep back the natives, who were beseeching in Hawaiian, vainly attempting to secure passage to see their friends and relatives on the island. It was a pitiful sight, — the dearest ties of life severed by imported disease, and Molokai, so near and yet so far, forever unvisited except by acquiring the dread malady. Their appeals, addressed to each officer in turn, could be met by nothing but the prompt refusal of a strict quarantine.

"Dr. Emerson, head and front of the arrangements, gave me my pass, — which stated that I went for scientific purposes, — and then we pushed on our way up the gang plank.

"The leper settlement had a great deal of interest to me medically, but I had become acquainted with the disease only through scanty textbooks. To me as much as to one of the laity it represented an unclean and unattractive malady, and although I had no fear of contagion, I anticipated that my sympathy would be strongly roused. Not knowing how the disease would appear, I felt

I should be glad to have the inspection of the first unfortunate over, so that I could study its effects in different stages without morbid interest.

"The steamer anchored at 5.30 A. M., about a quarter of a mile from the shore. As I looked from my stateroom window I could see lepers congregated on shore, and surrounded by saddled horses, ready for our service. The settlement, composed of neat white frame houses, looked more attractive than many coast towns of these islands.

"A band in white uniforms played characteristic native melodies, adding an element of melancholy which well suited the scene; for these people were trying to make the best of an opportunity afforded by the semi-annual tour of inspection. To them it meant a gala day, to us a sorry spectacle.

"After breakfast we were rowed ashore, and on reaching the wharf I caught my first glimpse of a leper. A small boy about twelve years old was comfortably seated on a rock. His face was rounded and enlarged, yet withered. His eyes, deep set beneath knotted eyebrows, and the nose (partly because the bones were destroyed, and partly from contrast with the swollen cheeks) looked almost lacking. His mouth, represented by a slit, was opened and shut when talking, in a peculiarly lifeless manner, and hypertrophied ears hung down like diminutive elephant's-ears. On drawing nearer I saw that his face was covered with tubercles varying in size from a pea to a bantam's egg, giving the appearance of a target for mischievous boys' putty balls. Eyebrows and eyelashes had fallen out; hands and feet were swollen, and the ring and little fingers of each hand had fallen off to their bases, while both great toes were bandaged as if in the same process of decay. Soon we were near enough to see similar characteristics in a hundred faces.

"On landing we walked to the so-called club-house, and while the officers of the Board of Health proceeded to business, the rest of us sat upon the porch, admired by a motley crowd of lepers, and entertained by the band, which played very well. It consists of ten musicians, some of whom belonged to the old Royal Band, and the leader still appeared in a cap with embroidered crown which he wore in his former proud position. He was a good-looking fellow, and bore no evidence of disease at this distance. All the rest were unmistakable lepers, and the man who played the flageolet was grotesquely horrible. Some of the instruments were fingered by hands which seemed too deformed to be useful. The bass horn was held by pressure of the arm against the body, as the player's left hand was so withered and drawn out of shape that it was useless; and as he had but two good fingers on his right hand, they had to be shifted in managing the three stops of the instrument. Another musician had lost an eye, and one limped as if his foot were nearly gone; while taken as a whole, the distorted faces gave a weird background to the performance.

"The assistant superintendent has been on the island as a leper for twenty years. He has the anæsthetic form, showing no tubercles or lost members; but his face was shiny, sunken, and like wax. When talking, his lower jaw dropped, and to close his mouth a distorted hand was pressed against the chin.

"The yard was packed with horses, and by nine o'clock lepers crowded amongst us, eagerly offering their horses for us to ride across the country, a distance of six miles. Doctor Emerson saw that I secured a good horse, and our party cantered away. It was a delightful ride, although each of us was on a leper's horse, in a leper's saddle, and handling the same reins that the diseased hands of a leper had handled; we forgot about it in the pleasure of the moment. Away above us rose a sheer precipice, and to the left lay the sea, making natural barriers shutting in the settlement.

"The Baldwin Home for boys is a neat little village, named for Mrs. Baldwin, who gave $5000 for its foundation. To this the Government has added $1000 and superintended the building of a pretty quadrangle. The frame dormitories accommodate eight or ten boys each, and in the centre of the square is a grass plot. The Government has great pride in the neatness of this home, and has spent much money in planting trees and shrubs about it.

"The authorities took pleasure in pointing out the comfortable arrangements, frequently stopping to indicate some of the worst cases, which all look more or less alike; but one young boy I shall never forget, with face so enlarged by tuberosities that his whole head appeared tremendous. His lips were so thickened and hardened as to make them from one to one and one half inches in thickness, and when they parted in talking the appearance was that of a wooden mechanism in action. The corners of his mouth became continuous with deep furrows in either cheek which made the mouth apparently of huge dimensions, extending into the middle of his cheeks. This with elongated ears and knotted face gave him an effect which I could liken to nothing human, but rather to a Chinese god of war. His small body corresponded poorly with the monstrous head and facial senility.

"Brother Dutton was introduced to me here, where he has made his home for the last eight years, in service of the lepers. He has done a great deal to make their lives happier and to arrange details of the home. Nobody knows what led him to take up this life, but it is reported by gossip that he was disappointed in love, or perhaps he committed some crime for which his conscience is now making him do penance. He is about forty years old, and his services are rendered without inducement or remuneration.

"Father Damien's tomb stands across the road beside the church he made with his own hands. He died of the disease contracted while ministering here.[1]

"On the way back we visited the crater of an extinct volcano, and reaching the club-house we found luncheon, sent ashore from the Iwalani. Then I strolled into the female quarters, only to find arrangements the neatest and most attractive on the island. This portion of the work is overseen by four Catholic sisters from Syracuse, New York, with their Mother Superior. Their handiwork is apparent in all the dormitories, and their influence in the figures of two young girls kneeling before the miniature altar of the chapel. The sister who guided me about responded very politely to my questions, and I could not but admire her quiet and attractive manner.

[1] Another monument in his honor sent by the Prince of Wales stands near the main landing.

"In one of the dormitories I found the only example of suffering which I saw at Molokai. The patient was middle-aged, her frame literally wasted to a skeleton. She had not long to live, and her labored breathing was exaggerated by a wheezing which comes when membranes of throat and nose are attacked; but a sorrier sight was her leper companion, who tried to support and fan her with crippled and bandaged hands.

"The Board of Health were busy all day. Twice a year they are compelled to examine all children born here. Those pronounced clean are taken away, if the parents wish it, but their consent is not always obtained.

"It was pitiful to see some of the young boys and girls on whom leprous spots were beginning to show, but to them it is only expected; and they have seen no other world than this. Mr. Mills and I went the rounds thoroughly, and as we had some time to spare took another horseback ride. Two lepers accompanied us on either side as guides.

"The settlement occupies six thousand acres of fertile land, where the large town of Kalaupapa was originally located, and includes the valley of Waikolu and the village of Kalawao. Surrounded by the sea on three sides, it is shut off on the fourth, toward the south, by cliffs two or three thousand feet high, — a beautiful spot which would do credit to a more attractive population.

"Rations are given out generously, and as no work is required, the lazy Hawaiian temperament is well suited. Their love of horses is gratified lavishly, for there are two horses for each man. To all outward appearance the lepers are better cared for than they would be at home; and as they have no fear of leprosy as a disease, and contract it by their own neglect and filthiness, they also gradually die without pain or worry.

"The painless character of this disease is certainly very fortunate. The first parts of the body attacked are the nerves; so that horrible deformities and loss of members surely accomplish their result, though with no discomfort, such as would be expected.

"I left the lepers of Molokai with less sympathy than I had anticipated; but as the band played our farewell, I was saddened by the thought of their failure to realize their miserable condition."

On this return trip of the Iwalani twelve "clean" children were brought back, who may perhaps entirely escape the fate of their parents.

"'Room for the leper! Room!' And as he came
The cry passed on, — ' Room for the leper!
Room!'
      And aside they stood,

> Matron and child, and pitiless manhood, all
> Who met him on his way, — and let him pass."

No such feeling as we have always connected with this horrible disease, and which poets and novelists have sometimes treated in ghastly fashion, troubles the Hawaiian native. His lack of dread is often the means of his contracting the disease. Transmitted largely through the saliva, all the members of a family, clean and unclean, continue to dip their fingers in the common bowl of *poi*. Unlike the white leprosy of Syria, this form is, thus far, equally incurable. Its germ has been found, and something may ultimately be discovered to neutralize or destroy it. Curiously enough, leprosy alone does not cause death, though death usually comes sooner to those so afflicted, because of its general weakening effect on all the organs, rendering them peculiarly liable to give way under slight strain from other diseases.

The Doctor returned almost without protest, during the progress of a farewell reception on the Coronet to some of the friends who had so lavishly entertained its company on shore.

The deck was draped with Hawaiian and American flags, and numberless pennants. Japanese lanterns hung thickly along the awning and among the green, while flowers and foliage filled every available spot. Cozy corners with cushions and rugs appeared unexpectedly here and there, the gig plied back and forth to the wharf bringing guests, and a native orchestra played softly through the enchanted evening. Supper and dancing, songs and friendliness until midnight; and then the quiet of a luminous tropical night, the Southern Cross dipping in the sea, the sweet life of the island a memory.

Only a busy morning remained before the long voyage. After luncheon, guests assembled for good-bys. Huge baskets of fruit, enormous stems of ripening bananas, flowers in countless bouquets and nameless luxuriant masses covered every spot, and a hundred *leis* were tossed over hats and shoulders of the departing company, until each prospective voyager resembled an animated tower of bloom.

Then with last farewells, a few lingering handshakes from deck to dock — native boys all about diving for dimes — we were off with dipping colors from the Adams, and final salutes reverberating. Lifting her white wings to the summer wind, out through reefs and breaking surf the Coronet took flight, over brilliant blue and green and purple water into deep-sea indigo beyond.

President Dole accompanied us for a few miles in his yacht, but when he had finally to turn back, there were more dippings and salutes, with the Williams yell for him, and the Amherst cheer for the expedition. Then the yachts parted too far for sound of word, while Tantalus and Punch Bowl and fair Diamond Head grew indistinct — yet more misty with atmospheric distance, and finally disappeared in gathering twilight. With full hearts we said *aloha* to these beautiful islands, already like home to each of us, with their friendly faces, their pathetic music, their gentle language like running water, their unsolved problems, and their brooding charm.

CHAPTER XIV

FOUR WEEKS AT SEA

> Alone, alone, all, all alone,
> Alone in a wide, wide sea.
> COLERIDGE, *Ancient Mariner.*

> Illusion dwells forever with the wave.
> EMERSON, *Sea-shore.*

WESTWARD and slightly south pointed the graceful bow of the Coronet, ever nearer the equator. A far southerly course would take fullest advantage of the regular trade-winds; but before they were entirely upon us the days were hot, quiet, tropically lovely, the glassy sea spreading white and dreamy to a misty horizon. Now and then a sunbeam struck through the prevailing haze from some far-off rift, and then a sparkling line, miles away, lay like silent surf breaking on an invisible shore.

With sea-water at 80° F., our days began by a plunge into the white bath-tank. Immediately after breakfast the awning was put up, impossible as it was to remain on deck a moment without it, in the heat and often blinding sunshine; and the great boom, swung far out over the water, was not shifted all day. Great was the heat, and the bananas, hanging in the shade, ripened apace — yet not too rapidly.

Always a surprise when mid-forenoon luncheon appeared, regular tiffin at one seemed but a few minutes later; when the afternoon had apparently but just begun, five o'clock tea was brought on deck — chased by dinner. And then came long, warm evenings under the brilliant stars. Occasional sunsets were fine, but as a rule not as gorgeous as on the Atlantic. Twilights grew shorter, darkness following quickly after sunset.

"One sun by day, by night ten thousand shine,"

but superb moonlight paled the glory of the Southern Cross rising higher above the horizon, the brilliant Scorpion, and all the tropic skies. Our nearest stellar neighbor, Alpha Centauri, became a distinct point of a new firmament. The nearest star! And yet so far away that its light, if starting toward us now for the first time, would not reach the earth for over four years. And proportionally our old friend Polaris sank toward the northern sea-line with his tethered constellations; even the tried and trusted Dipper descended alarmingly low, but at this season we never quite lost it.

Night after night, in the warm darkness, the infinite southern skies full of strange suns grew in impressiveness and solemnity. As Kipling says of the marvelous Indian stars, they seemed not "all pricked in on one plane," but preserved their own perspective through the velvet blackness.

Distances and difficulties are never insurmountable to the modern astronomer, with his splendid mechanical equipments. He questions the empyrean boldly, and little by little receives answer from illimitable space. Old observers contented themselves with studying motions and places of heavenly bodies; with long midnight vigils at their telescopes, and still longer computations, until every inhabitant of space that could be seen by aid of their instruments had, in addition to its own appointed path and position in the celestial vault, its corresponding place no less definitely in their columns of figures. But they knew nothing of what neighboring stars and planets are made; even the constitution of the sun was as a sealed book.

Now, the triumphant "new astronomy" lays its daring finger on the most distant stars, finding in Aldebaran and Betelgeux elemental substances not only identical with those closest to us on earth, but blazing as well in the majestic light of our own sun. Even the unformed nebulæ, ghostly tenants of cosmic space, perhaps birthplace of systems yet to be, have yielded part of their filmy secrets to the insistent spectroscope, and one by one the mysteries of the universe are unfolding to the keen eyes and trained skill of modern astronomy.

Once a galaxy of reticence, the chemistry of the stars is now known to be generically the same as that of the sun; and depths of space unsounded by the telescope are brought by celestial photography to eager eyes of waiting astronomers. A wonderful sensitiveness in photographic plates takes cognizance of faintest light from unknown suns blazing uncomprehended millions of miles away, which no merely optical telescope, however powerful, can show, and which encourages no present hope that human eyes will ever be able to visualize in future ages. The invisible is brought before us with irrefutable evidence; and distant wanderers through the stellar void which would otherwise have remained forever unseen are discovered, caught, and held for all time.

"Silent as death the awful spaces lie"

no less now than when to Immanuel Kant the starry heavens above were, with the moral law within, the most impressive concepts recognizable by the human mind. Warm foregrounds of villages and fields, mountains and forests, soften and make remote the solemnity of the nightly sky. At sea each soul seems alone with eternal verities.

Sundays were quiet days of blue and gold, morning service read in the saloon as before, well attended by the sailors; and long afternoons on the shaded deck, full of peace and liquid silence.

Our sailing master pursued the even tenor of his way, undisturbed by changes of crew forced upon him at every port. The first mate had joined the Coronet at San Francisco, a bluff man with a mighty voice, and not above seizing a halyard in his grasp of iron if he detected a bit of lazy hauling among the men; the second mate, a fair-haired Russian, reliable and resourceful, is now the Coronet's trusted first mate. The number of complicated knots which this amiable Andrew tried faithfully to teach some of us to tie, might have led to a profession in themselves.

Two quartermasters are charged with details on board more than the other sailors: they see that lights are in proper position, deck-chairs put away at night and arranged in the morning, the owner's "absent" flag and dinner-flags rightly hoisted in port, and altogether they are responsible for the minor etiquette of yachting.

Many new forecastle faces appeared on the trip outward from Honolulu. Several who came around the Horn in the Coronet had left at San Francisco, while others dropped off at Honolulu, — an uneasy class. The various names, however, seemed to remain always on board, and the Jims and Toms and Charlies were simply attached to different personalities.

One sturdy little sailor was hardly taller than the great wheel, seeming to command it with ease in spite of the momentary impression that it would take him in hand. Another was a typical stage-seaman, — young and handsome, with dark eyes and fine features, tall and well-formed as an athlete, with a throat like a strong white column; his bright and cheerful expression suggested just having finished, or readiness to begin some rollicking tenor solo before a waiting audience, about "joys of a sailor's life, yo-ho." But he never did. Mother Goose's Simple Simon daily helped to set sails, or scrub decks, or took his turn at the wheel. When hauling on a halyard he put in his whole soul, with facial results appropriate to the instant of committing a complicated murder. The same sailor who spun great yarns remained on board, and his stories grew as the voyage progressed. One day the Doctor came aft, from an excursion to the bow, and related a surgical tale deserving record. Big Jim was apt to regard the profession of his auditors.

"I was once thirteen months in a Bombay hospital," he announced, "and at the end of that time the doctors had to take six inches out of my backbone. So I never grew any more. Stunted for life."

This story only failed of its best impressiveness because the victim was the tallest sailor on board. Having warned some one reading on the forward deck against such dangerous employment, Big Jim said, "Why, I used to read all the time myself, and it made my eyes so bad I had to go to a hospital and have 'em taken out and scrubbed. The doctors found they couldn't do it well enough there, so they sent 'em away to be cleaned, and I didn't get 'em back for three months."

Days grew constantly hotter, a bird now and then forming the chief incident in a wide sky, although whales occasionally spouted or sharks darted through the water, their sharp fins easily recognizable; once a series of fine water-spouts swept our early morning horizon. No sails appeared. If there were "ships that pass in the night," they remained invisible.

But winds were at last with us, steady and strong, and good runs were made, — the whole voyage beautiful enough to last forever without protest. Scrapbooks were brought up to date, even to the *aloha* from Honolulu; journals and letters flourished, chess-players became finished experts, decorations (in the shape of various pennants) were painted in the saloon, serious work of the expedition progressed, and days flew by on noiseless wings. The Mechanician, surrounded by wires, batteries, tools of all sorts, and small boxes of deft devices, sat on deck with head bent forward, ardent spectacles gleaming, as he toiled early and late at the inventions of the Astronomer, who was occupied near by in making the calculations necessary for experiments with different exposures in all the twenty or more photographic instruments, — each being arranged to take that automatic series of pictures of its own already described.

Occasionally the Doctor brought forth cases of shining and suggestively ingenious tools of another trade, newly purchased for this expedition, and all in best of condition for any dire calamity. Happily lack of specific use necessitated much attention and polishing to avoid sea-rust. When free from one sort of paraphernalia, both deck and big table below were generally strewn with the implements of some other profession, in orderly confusion. Sometimes they were summarily swept away by Alfred "the Great" as meal-time approached, proper serving at the expected moment being a far more serious consideration than any mere eclipse, or celestial streamer.

As for a little Richard barometer in the companion-way, it was an intimate friend of all, a glance at its telltale cylinder being an invariable but half-unconscious incident of every trip below, if a dozen times in a morning.

The Captain barely escaped the loss overboard of his birthday at the one hundred and eightieth meridian, but its rescue was celebrated by a huge cake with candles, and many gifts unsealed from home. Delightful contralto, bass, or tenor solos diversified those evenings when temperature would admit staying below with the piano; or quartettes on deck floated over lonely Pacific wastes which may never again stir those solitudes. Chief, too, developed still another talent, giving us burlesque operas, accompanied by the guitar or autoharp in thrilling style, some of his final trills and cadenzas falling little short of the sublime, as he dramatically bewailed a broken heart in brilliant falsetto.

And still Polaris sank lower, the Cross riding nightly higher in our southern heavens.

Hoisting the main topmast-staysail was always a pretty sight. When lowered and stowed away it is delicately tied together in a long roll, and hauled into position still tied; but when in place, the wind and a slight jerk breaking the little cords in speedy succession, it falls apart white and graceful, and is quickly made fast.

At early morning, oftentimes, a curious noise like the rush of an amateur cyclone sounded over our heads, but it was only a sailor scrubbing his white duck clothes on deck in sea-water with a big brush and salt-water soap. In the main they were fresh-faced, wholesome men, these sailors of the Pacific, quiet and industrious, with great pride in the beautiful Coronet.

The shanties still continued, our mate, as on the previous voyage, singing the solos, and a hearty chorus aiding greatly in hoisting the mainsail.

"Oh, Bony was a warrior" seemed a favorite:

1. Oh, Bony was a warrior, wa, a, wa-a, Oh, Bony was a warrior, wa-a — John French war (Jean François.).
2. He was a holy terrier.
3. Oh, Bony went to sea one day,
4. He went across to Eng-land,
5. The England did a' stop him.
6. Oh, Bony went to Moscow,
7. He gained a bunch of roses there.
8. Oh, Bony went to France again.
9. The England went a' after him,
10. He brought him Saint Helena,

where his adventures seemed to lapse. "Whiskey boys, whiskey," was no less popular.

"A long time ago" was fitted with words describing the escapades of a certain sailor at Honolulu who had boasted of his income from his real estate in that city: —

"It was the merry month of May; Wa, wa, wa, wa.
The Coronet at Honolulu lay, A long time ago.

CAPTAIN AND OWNER OF THE CORONET

"Jimmy went on shore that day, etc.
To draw his rent and three months' pay:
Jimmy did not come back that day"—

and so on through a long tale varied as feeling toward Jimmy rose or fell.

After the course was changed to northwest, winds became curiously fitful, almost as if the edge of a typhoon had passed by, so abnormal were the conditions. Showers fell, general rolling prevailed, winds died out, or else sharp breezes sprung up from unexpected quarters. For several days anything was anticipated, but one afternoon a regular wind began once more, after a heavy rain; coolness and comfort returned, and ten knots were easily made. A high gray sea was running, though the water still showed a temperature of 80° F. Then a rollicking blue morning with sparkling white-caps, and everything was natural again.

Sextants and other nautical instruments abounded to an unusual degree, and observations were not confined either to Captain or to noonday sights—but Polaris, Spica and Antares were watched at night, by astronomer as well as yachtsman. Enough solid navigation to direct a fleet was carried, boxed in its own little mahogany nests.

Visitors on board were rarer than on the way to Honolulu; one exquisite little creature like a poeticized corona was caught,—a delicate blue centre with a double row of lighter blue encircling rays. Twilights once more grew long.

A day or two before the coast of Japan should have been sighted, flocks of birds appeared, the breeze suddenly increased to sixty miles an hour, while huge gray rollers again broke all over the tossing sea in sharp white foam. Yet the wind was in an opposite quarter from its normal direction, if indeed this disturbance were the edge of a typhoon sweeping up the coast. Quick orders for lowering sail rang out; in the confusion of tramping feet above, and the booming wind, all sorts of expressions came down the companion-way, cut into bits in their descent, and fraught with mysterious import. "Clew up your topsails," "Let go your throat," mingled with directions about the lee lift and the main sheets. I listened in vain, however, for my favorite order on board, "Jig up your peak." To-day's crisis demanded quite the opposite of "jigging up" anything. But in an incredibly short time phrases were translated into an accomplished shortening of all sail. Nothing remained but the main trysail and a jib; it rained with tropical lavishness, and once more we were "hove to" near the coast in a wild swirl of waters.

And still the Pacific had retrieved its character. Since 1887 I had felt it entirely misnamed;

that its fog, high seas, and general roughness of demeanor demanded an apology, at least an explanation, from those dead and gone worthies who saw fit to call it Pacific. But probably they had not sailed a great-circle course from Vancouver, as our previous expedition did. Now, after traversing its enchanting southern water spaces, with day after day of shining sea, and tradewinds, with no necessity for racks on the table or "fences" at night — these things quite obliterated all memory of that other unfriendly northern Pacific which in 1887 had treated the old Abyssinia so unceremoniously. Except for a day or two, this voyage had been a tropical harmony in blue and gold.

And after this one tempestuous night, the morning dawned fair and lovely, but greeted no longer by a sapphire sea to reflect the brilliant sky. The Coronet was unmistakably in the *Kurosiwa*, the "black current" of Japan; the water was dark green, and full of drifting sea-weed.

Before sunset of that bright Sunday, the twenty-first of June, two or three islands appeared, — Mikura, Miaki, and Vries. Then the incomparable cone of Fuji lifted itself against the sky — that well remembered landmark which was our last sight of the beautiful land nine years before, and without which Japan could not be Japan. Faint and far away, but unmistakable, and as fair as when, the morning after its miraculous creation, this "new born child of the gods" caused the sailors at sea to rub their eyes and wonder if it were the *Iwakura*, eternal throne of heaven, come down to rest on earth out of the many piled white clouds above. The majestic cone vouchsafed royal welcome, though less clear than at his gracious dismissal.

And then a fishing-boat or two appeared,— first sign of human life other than our own in all the four weeks' wide stretch of lonely sea. As darkness came on, great Fuji melted from sight, and here and there torches twinkled unsteadily from fishermen setting trawls. The Captain remained on deck all night, and his guests went below with mingled sensations of memory and anticipation.

CHAPTER XV

JAPAN REVISITED

> Thank God for tea! What would the world do without tea! How did it exist? I am glad I was not born before tea.
> SYDNEY SMITH, *Memoir*, i. 383.

DANGER of disenchantment lurks about a return to distant lands whose memory has been for years enshrouded in rosy atmosphere. The halo idealizing our recollections down the vista of years may dissipate into nothingness once the actual comes again in sight.

Will the air be as sweet as in those dreamy retrospects? Will the beauty be as all-pervasive, the charm as haunting? All the mistily bright June morning when the Coronet was beating up Yeddo Bay between green shores on either side, this unspoken wonder seemed to hover half-unconsciously in the sunny air.

For nine years the name of Japan had recalled pictures of dainty little women thronging its streets in bright dresses and gay parasols, and of shops full of fine old swords and other relics of samurai days, sold for a. trifle, as being of no farther practical value in the modern life then beginning to overwhelm the beautiful land.

Memories, too, of *jinrikisha* rides through quaint streets, when the coolies pulling the fascinating little carriages had known scarcely a word of English, and were more than satisfied with ten *sen* an hour for their exertions, and of night rides when the shops and open booths were lighted by flaring torches, and foot travelers and jinrikisha bore their own painted lanterns swinging in the soft darkness; of happy babies strapped on the backs of sisters or mothers, to spend long days in utter content which excluded even the knowledge of how to cry — all these thoughts of years, and countless others, were concentrated in one bewildering mental retrospect, as we sailed once more up the lovely bay, in the era of Meiji 29.

Familiar places came into view one after another, the sharp promontory guarding Mississippi Bay, then the houses on the Bluff nearly hidden in verdure; farther on the mercantile parts of Yokohama, and the Bund with straggling pines on the water-side, low houses facing the bay behind verandas and garden-walls on the other; great Fuji dimly brooding over all, unchanged against the sky, — and we were once more casting anchor among the men-of-war of all nations, inside the superb new breakwater.

A few years ago no barrier had raised itself against the tempestuous seas which almost at a moment's notice often turned the harbor into a boiling, seething mass of tossing waves; when it was impossible to induce sampan or even steam-launch to take one out from shore, even if an already promised tiffin or dinner on one of the men-of-war were involved. Now the harbor is a safe and quiet anchorage.

Before the Coronet actually came to rest a dozen sampans had surrounded her, their wooden anchors lying in the bow as of old, and propelled in the familiar way by one huge oar at the stern, but no longer wielded by what had once appeared animated bronze statues. Instead, all were decorously clothed in dark blue cotton garments, or attempts at European array, although the big, picturesque hats still prevailed.

But English was actually spoken by the men who held up cheap porcelain and coarse cloisonné for sale from the native boats gathered about. Rather imperfect, but generally definite "American," it was still successful as to import. The only chance for that class of wares with foreigners is immediately upon arrival, when everything seems beautiful. The discriminating faculty of the traveler soon comes to the front, and he speedily becomes critical in all matters of Japanese art.

Rather surprised, even their *savoir faire* somewhat upset by the few sentences in their own tongue tossed over to them, relative to price and quality of their wares, these light-hearted vendors of unattractive articles paddled away; and hotel-runners, provision dealers, laundry-men, and every variety of tradesman clamored in their stead. But quarter-masters and stewards kept the yacht decks free from the amphibious host.

Reporters were by no means left behind on the American shore. Delightful little gentlemen, some in *kimono* and *obi*, English boots and Derby hat; some in paper or celluloid collars, crowning elegance of a limp suit of pongee silk, or seersucker; others in the beautiful native dress unadulterated, — all were still the same deeply bowing, smiling, spectacled, courteous class we remembered. One of these gentlemen prepared for his *shimbun* (newspaper) a serial upon the expedition and its adventures which ran through four numbers. And another came on board with the startling announcement, very calmly made, that he had "come to take the life of chief of expedition — for Japanese news-paper."

Remarkable disturbances of nature seemed to accompany the Amherst Eclipse Expedition upon its travels, and the first news heard by the voyagers, quite starved for information as to what had been happening to any of the earth's inhabitants during the last month, was intelligence of a terrible misfortune in northern districts of the main island. It was learned that an enormous tidal wave had within a few days devastated more than thirty towns, washing away nearly six thousand houses, and destroying between thirty and forty thousand persons. Since the great earthquake in which Yeddo (now Tokyo) was nearly swallowed up, forty years ago, Japan has had no such calamity, not even the Bandaisan eruption of 1888, or the Nagoya earthquake of 1892.

Detailed accounts of this appalling disaster were still hard to obtain, for the few survivors in the devasted districts were too dazed to give clear descriptions of the horror which befell them. But it was known that a seismic wave, some persons declared one hundred feet in height, the majority uniting upon an altitude of about eighty feet, swept across the land with irresistible force. Along a coast line of one hundred and seventy-five miles in one province alone, the seaboard of three districts was overwhelmed, — Miyagi, Iwate, and Aomori, extending from Hachinoye on the north to Kinkasan, an island at the mouth of the bay of Sendai, on the south. Several shocks of earthquake were felt during the few hours preceding, and shortly before eight o'clock in the evening of the 15th of June a terrifying noise was heard, like the boom of gigantic artillery, — the simultaneous firing of hundreds of cannon; a black wall of water was seen advancing from the ocean with fearful speed, and in less than two minutes whole towns were swept away and thousands of human beings perished in the onward rush of this watery monster, and there were not survivors enough within reach to bury the dead who had not been sucked out to sea by the retreating tide.

Their Imperial Majesties, the Emperor and Empress, came at once and nobly to the rescue, the Mitsui family contributed scarcely less, and the Tokyo journals opened subscriptions for relief of starving survivors, the "Jiji" collecting in a few days over ten thousand yen, and the "Nichi-Nichi" more than eight thousand, while the Iwate branch of the Japan Red Cross Society established temporary hospitals among the suffering people. The number of victims was at first greatly underrated.

All the habits, even the methods of thought, return in plunging once more into a land as distinctly foreign as Japan, and even in Yokohama, where strangers from other countries most abound, the native atmosphere is hardly adulterated enough to change the general effect. The sunlight, the tints and odors of years before, were unaltered. Even jinrikisha riding had lost none of its charm, even if the runners, now speaking considerable English, did show a growing affinity for their far-away fraternity, the cabmen of American cities, in demanding whatever they could get for fares.

The effort to adapt manners and customs to an imported standard, redoubled since the brilliant termination of their war with China, was everywhere apparent.

Many signs are displayed in our own familiar letters, instead of the picturesque floating strips of dark blue cotton with their decorative ideographs in white. Still, these are not superseded, even in Yokohama, and the streets are like one long holiday parade, — they actually throb with mysterious vitality, the ideographs quiver with meaning; a vivid picture comes before the mind with each character floating in the wind. To a Japanese "it lives, it speaks, it gesticulates."

Art is in the air, until suddenly one comes across an English sign, perhaps after this style: "Dealer in of fan circular fan umbrella."

A fierce and sturdy-looking individual in abundant whiskers and Americanized dress stands painted guard over one shop, with a high boot on one leg, a shoe on the other foot. This not being quite definite enough, the legend runs "Shoe to make to form."

"Watch and gold silver ware belonging" was quite clear, also "Drinks and courserues."

A well-known dealer in curios advertises "Our shop is best and obliging worker that have everybody known, . . . We can works how much difficult Job with lowest price insure, please try, once try don't forget name Whisky." Possessives and plurals, in truth small matters, are considered rather too trivial for use. Possessives, indeed, become expensive if one is telegraphing in English, each adding several *sen* to the sum total of charges.

"Wholesale and retail seller shop," and "Landing, shipping customs goods forwarded to parts" were easy to comprehend, as well as "Transportation of several goods and baggages of steamboat and railroad," and "Wine beer and other." But "Do you love your life or rather" was more of a conundrum. An odd combination profession seemed to be implied in "Portrait painter and dealer in Manila cigars," while another shopkeeper announced above his entrance "glass and lumps" as his stock in trade. A sign of rather startling import asserted that within might be found "Lamb, corpses and provisions in seasonable rates." But we purchased our chops elsewhere. A collection of foreign signs during this transition period of Japanese advancement would show new possibilities in the English language.

Within the shops more articles were obviously

made for travelers from other lands, — sure beginning of art degeneration in any country.

Time had stolen little in nine years from the two famous sisters Tanabe-san and Kin-san. Once beautiful as well as fascinating, they still remain exceedingly attractive. Their uncle, in power in the province or ken in 1859, signed the articles of treaty with Commodore Perry, and the two charming women have always lived in an atmosphere of the world at large, while yet preserving the dainty sweetness of their race. Acquaintance with them is a definite, integral part of Japanese experience; both sisters speak no less easily in French, German, and even Russian, than in the English which they use so prettily, and the little silk-shop where embroideries may be bought accompanied by gentle compliments in English, manners to credit the graceful *régime* of old, and pale yellow tea of delectable flavor, was still pleasant meeting-ground for many nationalities.

In the celebrated tea house at the top of the Hundred Steps, Kin-san preserves many mementos of her ancestors and Commodore Perry, as well as an interesting guest book, in which may be found the cards of hundreds of travelers of distinction visiting Japan during twenty years. Here Kin-san dispenses cosmopolitan hospitality, filtered through customary Japanese forms, and Yokohama will lose one of its great delights when she ceases to serve tea and sweetmeats from her lofty veranda at Fujita, almost overhanging the gray tiled roofs of the city far below; and when her soft voice shall no longer accompany her elaborate playing on koto and samisen.

One of the most picturesque spots in Yokohama, Fujita is reached either by literally toiling up a hundred steps from the street below, where the *kurumaya*, trusting foreign ignorance of locality, will basely leave a confiding fare if he can be so imposed upon; or by a winding road ascending in easy stages to the rear of the teahouse. At night the view is a sea of twinkling lights below. Foreigners have always played a large part in the experience of both these dainty women, whom necessity compelled to transact business for themselves; and without ever leaving Japan they have seen the world in very attractive guise.

Certain distinctive habits have by no means been outgrown in all the incoming rush of modern ways. In making *kimono*, for instance, different sorts of stitches having reference to the prospective wearer were still used, a system perhaps a little less elaborate than the Morse telegraphic code. A long and two short stitches, one short and two long — these combined in a variety of ways indicate that the garment is for a man, or a married woman, a young girl, or a child, or perhaps for a girl about to be married.

And still the pleasant life of foreign residents went on much as it had years before, open port life showing fewer changes than more purely native places. Our old friend, consul-general in 1887, during the first Cleveland administration, was no longer there, but instead in Korea. Others, however, were still at hand to give friendly and well-remembered greeting; and among the officers of the men-of-war there were, as always, many acquaintances.

The Coronet's next neighbor in harbor was the Olympia, flag-ship of our Asiatic squadron, now more famous from the great Manila victory. The Detroit lay peacefully near by, and the French cruiser L'Alger, while English men-of-war and even a Mexican brightened the bay, with a number of Japanese merchant and naval vessels. Several small yachts skimmed lightly about, or lay at anchor near the Bund, and daily in landing we passed a schooner yacht but just returned from the South Sea Islands. A pretty craft, apparently manned by one huge Fiji Islander, she took little part in the gay harbor life flashing around her on all sides.

At sea one is never allowed to forget the passage of time, for two, four, eight bells are always sounding, the hours and half hours chasing each other in a mad rush for eternity; here in harbor there was little danger of wondering what real time actually might be, although each nation, and almost each vessel seemed to have its own notion of when the bells should be struck. Only a few seconds apart, they formed a pleasant chiming all over the bay, clear and loud, or soft and distant, echoing from one to the other in melodious iteration.

At colors every morning a fine concert from the Olympia always greeted us. When all the ensigns and pennants slowly ascended into place at eight o'clock, and as our company, if on deck so soon, stood with bared heads while the Coronet's stars and stripes went up, the flagship band played America; then the Japanese national air, — a curiously characteristic melody, — followed sometimes by the national anthems of the other countries represented by the men-of-war lying in harbor, and ending always with Nancy Lee, in pleasant compliment to their little neighbor the Coronet, whose especial song it is. There were bands, too, on some of the other men-of-war, and bugles playing familiar calls. Talking from one to the other by day with different signal flags gayly floating in summer breezes, and evening conversations by flashing colored electric lights, made harbor life vivid and picturesque.

Naval hospitalities flowed in upon us, — a tiffin by the Admiral, dinners by the Ward Room officers, dances, teas; while the pretty Coronet held her own bravely in the exchange of social courtesies; and on shore were no fewer festive occasions.

One memorable evening ceremonial tea, *cha-no-you*, was served for our benefit.

In all the modern rush of nineteenth-century life, beautiful old customs will be in danger of dying out, or at least of being pushed from sight. On the previous visit so many more of the purely historic, hereditary and traditional forms were practiced than seemed available this time, that it was a delight to see once more the elaborate tea-ceremony in all its solemn impressiveness.

Young girls are trained a long time for presiding at this function, and every motion is adjusted in accordance with deeply philosophical and ancient usage. The ceremony itself and its underlying principles have been so often and minutely described that I shall but refer superficially to the features which were most apparent to the Coronet company, sitting in a circle on the floor in waiting silence. The tea itself is a choice and very fine green powder, every implement old and valuable, heirlooms if possible, and kept especially for these occasions.

When a cup of this exceedingly delicious beverage is set before the guests, in order of their rank, each lifts it slowly to his forehead, after bowing low, turns it ninety degrees counter-clockwise, and drinks it with deliberation, so regulating his sips that three and one-half will just exhaust the contents of the cup, the last being taken with a slight indrawn hiss to express intensity of appreciation and pleasure. Between sips the cup is gracefully shaken, also in a particular way, to stir the powder at the bottom. The finger should wipe the edge of the cup where one's lips have touched it, the finger itself wiped upon a little piece of soft, once-folded paper already laid upon the mat. Another piece, folded in a point, holds a sweetmeat afterward to be taken home. Later, the tea-caddy with its fragrant green powder is passed from one to another, for admiring scrutiny, also expressed in a special manner; as well as the long-handled ivory spoon with which the powder is transferred to the teapot. It is all very slow and stately and cultured.

"Well," remarked one of the guests, straightening his American back as he emerged from the dainty dwelling, and started for the Bund, "It does n't take long to stay a good while sometimes."

But truly lovely was the return to all the grace and culture, the exquisite breeding, the constant thought for the happiness of others — the artistic life of Japan. Even sitting on the floor has its glamour, if one takes the right mental as well as physical attitude, and the genial sweetness of the entire country is so pervasive that the best of one's nature expands unfailingly in its sunny atmosphere.

Japan is changing, and noticeably; but Chinese compradores continue to walk unsmilingly through the streets in quiet majesty, and many years must pass before expressions of the national spirit will fall naturally into the commonplace ruts of other civilizations. The past had perhaps been canonized, and the present was different, but there was no disenchantment. The old-time charm exerted its spell as before, although a few babies were heard crying with truly western vehemence, evidently the result of foreign influence, and at a delightful tea-house entertainment one evening no painted lanterns swung in the breeze, but electric lights flashed forth from a bronze chandelier.

Tidal waves destroying lives and temples and monuments in Japan are not a modern innovation. From earliest times the country has been occasionally overwhelmed by various forms of destruction. Floods swell the rivers, wash out railroads and drown rice-fields; earthquakes wreck whole towns, volcanoes bury provinces.

Yet always energetic, hopeful, aspiring, the Japanese take fresh courage from misfortune and rise to renewed power and mastery in the life of the far East.

The brilliantly successful issue of the war with China has given the Japanese new faith in themselves, and the spirit of modern progress is abroad in the land. As a nation they will ultimately incorporate whatever is best in our civilization with their traditions, hoary with centuries, beside which the short history of America seems but an episode. If only they are far-seeing enough to retain what is best and most characteristic in their own civilization as well, the combination will make a country of modern enterprise, coupled with the artistic bequest of ages, which the world has never seen.

JAPANESE NATIONAL AIR

CHAPTER XVI

DEPARTURE OF THE EXPEDITION

<blockquote>
In this world, where civilization grows at the expense of the picturesque, it is something to see a culture that knows how least to mar.

PERCIVAL LOWELL.

Tell us of thy food, — those half-marine refections,
Crinoids on the shell, and Brachiopods *au naturel!*

BRET HARTE.
</blockquote>

YOKOHAMA'S native quarter was still like some animated fan or screen. Wooden clogs (*geta*) clicked well-remembered music, and the little teapot ladies of the thoroughfares made no visible attempts to sport in European dress. That fad seemed to have died a natural death, and attractiveness in street scenes is apparently assured. The pretty *kimono* were out in full force, with all their bright colorings, the flowing sleeves doing service as capacious pockets for paper handkerchiefs. Young girls in scarlet underskirts still clattered along with tiny, black-eyed brothers swinging on their backs; gay sashes (*obi*) and elaborately dressed hair gleamed in the sun, and bridges with their crowded passers were more picturesque than a picture. And however poor or low in caste a Japanese woman may be, she seems never too ignorant to keep her hair smooth and shining as a matter of course. One meets no fuzzy, rough-haired girls, in any quarter. Unhappily the pretty little women had abjured the gay, many-ribbed parasols, appearing in their jinrikisha sedately shaded by black silk umbrellas of very ordinary shape. But children still flourished the brilliant paper ones.

The Astronomer was at once busy with Government officials, with the Imperial Weather Bureau, and with our own representatives at the Legation, and very soon the station for observing the eclipse had been selected.

A new system of meteorological observations made before an eclipse and with special reference to it was inaugurated by Professor Todd in 1890, for that of 1893. Taking the exact track as soon as published in the Nautical Almanac, and having careful observations made at the best and most accessible points, gave excellent results on that occasion. As the Ephemeris is issued about four years in advance, this insures three complete series before an eclipse. Noting the general meteorological conditions of the heavens is not sufficient, for the sun is in a particular part of the sky at a given hour, so that the observations must be of special character, and with distinct reference to the position of the sun, season of the year, and hour of the day when the eclipse takes place. The tabulation of this information assists greatly in selecting the best stations for eclipse-observation, and those who followed the indications as to clearness of sky in 1893 achieved the best results.

In that year Professor Todd wrote to the Director of the Imperial Japanese Weather Service, requesting observations at different points in the Hokkaido, and his suggestions were carried out in every particular. Professor Nakamura, of the Central Meteorological Observatory at Tokyo, had printed and distributed to the different legations a pamphlet for the information of eclipse students, containing not only all the observations referred to, but a sufficiently minute discussion of them to enable all the astronomers to weigh most intelligently the probable chances of clear skies at every available point in the path of totality. The establishment of any station is thus made with full knowledge of whether it is best or worst in probable clearness; and if obliged to plant himself in some less hopeful location, the intending observer takes his own risk, with eyes wide open to the law of probabilities.

The three provinces of Yezo in which the shadow fell were Kushiro, Kitami, and Nemuro, each containing several towns, most of them small and but little known to foreigners. During 1893, 1894, and 1895, tri-daily observations were made from July 25 to August 25 at two o'clock, half after two, and three o'clock, at a number of these villages in the eclipse track, the results being carefully collated in comparative tables. From the percentage of cloud at the observation hour itself, Akkeshi, on the southeast coast, came first in probabilities of clearness, and Esashi, on the northeast coast, second; but from the point of its constancy thirty minutes before and after the eclipse, Esashi presided over all the others, as shown by the full tables given for thirty-two days at seven towns.

The selection of a station always involves much care and forethought, and responsibility enough to whiten the hair of any one except a philosophic astronomer, accustomed to take chances with nature. The probabilities at Esashi were considerably more than half in favor of clearness, and after studying the reports and tables carefully and consulting with the meteorologists, the Professor finally selected that point as his observing station, although it is farther and more difficult of access than Akkeshi, of which he had thought before our arrival in Japan as a probable location.

And so Esashi, eleven hundred miles north of Yokohama, became the scientific Mecca toward which these pilgrims would wend their way, — a

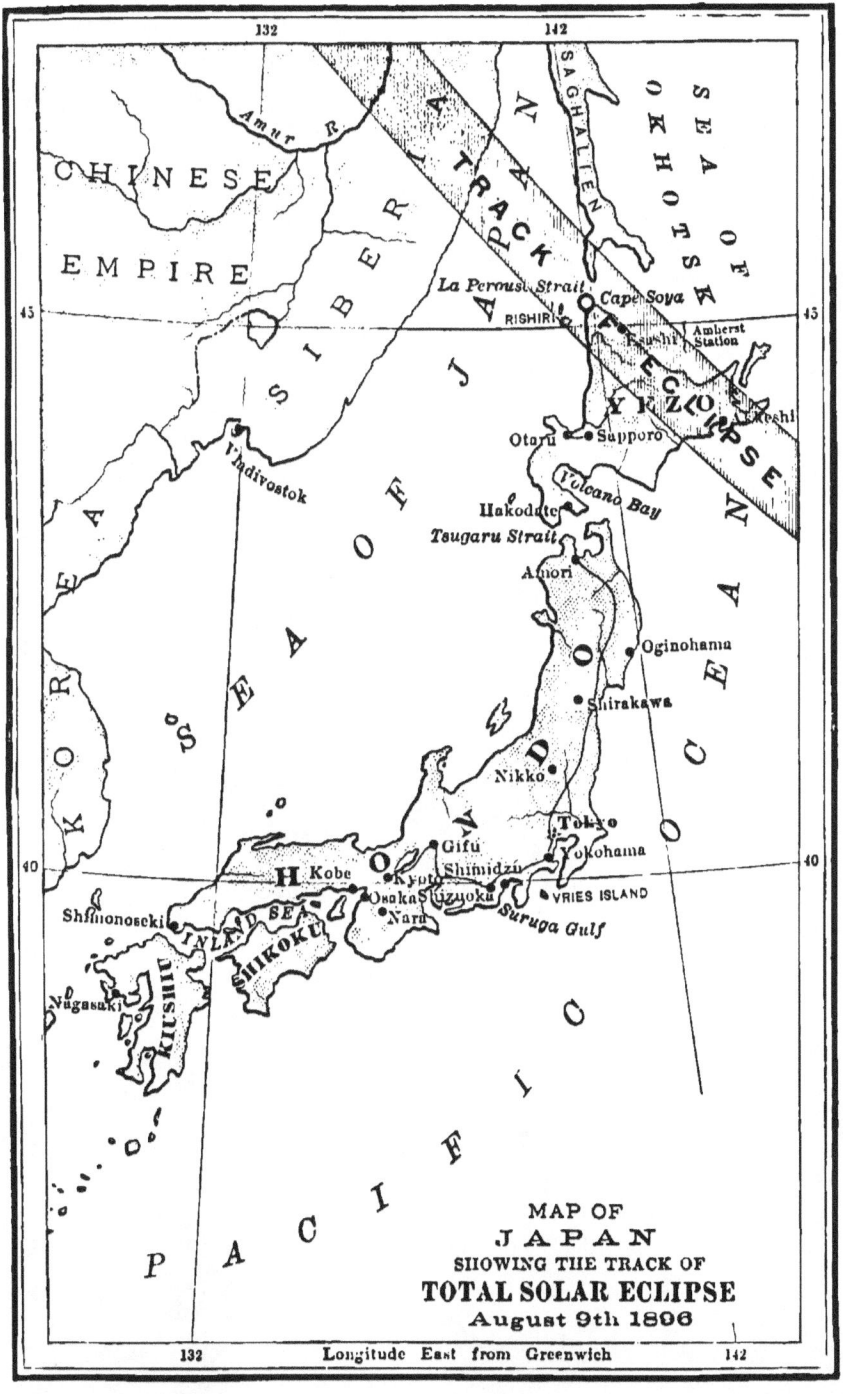

region so remote that native steamers had but recently begun to go there at all, and that infrequently; a village on the shores of the Sea of Okhotsk, among the hairy Ainu, the aborigines of Japan, cut off by many hundred miles of impassable forest and mountain from even Sapporo and Hakodate. At least the prospect for unusual experiences looked hopeful.

The distance of Esashi in Kitami from Yokohama, and the sort of coasting voyage necessary to reach it, put out of question all possibility of sailing there in the Coronet; so it was planned that she should convey the unscientific contingent southward to Kobe instead, making afterward a trip through the Inland Sea.

But the imperial government was most courteous to the expedition, affording every facility possible, which included, with truly royal generosity, requests to both railroad and steamer corporations for free transportation for the whole party and the instruments to any point they might select, and many other favors which greatly enhanced our comfort and convenience.

Official matters move slowly in Japan, and our imposing array of introductions and documents from Washington needed time for fullest availability. The Astronomer wished, if possible, to leave Yokohama not later than the first of July.

The interest of scientific men in this eclipse

was shown by the large number assembled for its observation — French, English, American, Japanese. France was represented by M. Henri Deslandres, then of the Paris Observatory, accompanied by M. Millochau and the brothers F. and J. Mittau. He also chose Esashi, whither the French cruiser L'Alger was soon to convey him from Yokohama with his fine equipment of spectroscopes. Genuinely modest as he is, Professor Deslandres no doubt hoped to bring back from Kitami wilds some solar discovery not less significant than his trophy from the African eclipse of 1893, — the rotation of the corona with the Sun.

Professor Schaeberle, head of the Lick Observatory expedition, stationed himself at Akkeshi with his party, consisting of Mr. Burckhalter of the Chabot Observatory at Oakland, and others. Professor Terao, Director of the Tokyo Observatory, chose Esashi; and the Astronomer Royal of England, Mr. Christie, with Professor Turner of Oxford, and Captain Hills, of the Royal Engineers, soon arrived by a Canadian Pacific steamer, and proceeded forthwith to Akkeshi, in Her Majesty's ship Volage. But a few days remained before our own expedition would depart northward — days filled with hurry of preparation, yet leaving time for enjoyment of many native and foreign courtesies.

A certain half-tropical, gently pungent odor, not precisely that of flowers, or luxuriant vegetation, incense or moist atmosphere, old embroideries, or the culture of ages, but perhaps all of these together, and more powerful to awaken association than even sight or sound, haunts Japan like a spiritual aroma. With its first dimly suggestive breath the nine years' chasm was so bridged that life might almost have gone on ever since without a break, in this dear, dreamy lotus-land.

An old friend, and graduate of the college represented by the expedition, whose father, at one time highest in command in the Imperial Navy was just then Governor of Formosa, invited us for an evening of Old Japan at the Maple Club in Tokyo. A resort of the nobles, fine entertainments are constantly given there, under exquisitely characteristic conditions.

Before the dinner, which began about six o'clock, we drove with our friend to the Imperial Gardens by the sea, — a charming spot, not accessible to the public, and laid out with that taste and skill peculiar to the Japanese landscape gardener. Advantage has been taken of its situation to introduce many beautiful water-ways directly from the bay. When these winding inlets are to be crossed, the bridges do not go uncompromisingly over from one bank to the other, but abound in unexpected corners and turns and "jogs." And the railings are not of plain, square timber held together by heavy nails; instead, a light and graceful bamboo rail — its fastenings of fine wrought iron, each a work of art in itself. A daintily decorated tea-house awaited the guests, and servants in livery of the nobles explained, and guided them to the finest points.

Fish, apparently afflicted with hysterics, leaped constantly from the water, often two or three feet above its surface. Ingeniously dwarfed trees stood here and there; superb forest trees as well, while delicate maples, with their seven-pointed leaves, cast lovely shade in the summer day. Finding an eight-lobed maple leaf is as desirable as a four-leafed clover in another land.

From a hill was gained a view of the blue bay with its fleet of square sails gathered into stripes after the old, well-remembered fashion.

There is no expectation that guests will not stroll anywhere over the fine turf, but wherever these wandering footsteps are liable to converge — as perhaps here and there at a few moss-grown stone steps — a gravel path begins suddenly in the grass before the steps are reached. Sometimes but a short bit of path is required, and there is nowhere that appearance of unintentional footwearing often marring constantly trodden parts of lawn.

At the Maple Club entrance a bevy of pretty maids welcomed the guests, taking their shoes before conducting them through a long series of polished corridors to an airy room open on two sides to an elaborate garden, a large lotus-pond just beyond, and great Fuji eighty miles away, against the brightness of a sunset pageant.

Here the Countess, mother of our friend, a sweet-faced lady in gray silk *kimono*, met us with warm greeting, though herself speaking no English. Two young girls, daintily dressed in native costume, with superb *obi*, also bade us welcome in friendly Japanese. Their melodious names were found, upon interpretation, to mean something about a flower and the shining of a red star.

Our host, who had been married since his return to Japan, then presented his small daughter, brought in her nurse's arms for a few moments, a gorgeous baby of six months arrayed in magnificent gold brocades reaching quite to the floor, the expression upon her little face peculiarly intelligent and high-bred. One could hardly have imagined her as belonging to the same race with the chubby-cheeked, fringe-haired *akambo* everywhere seen tossing about on many a back.

Among the guests were a young viscount, also an old friend, who with his father and mother and beautiful sisters had on our former visit paid us many delightful attentions, not the least of which were gifts of memorable embroideries; and a professor in the Imperial University, a graduate of Amherst, to whom we owed many a pleasant memory of 1887. As the three Japanese gentlemen spoke perfect English, social intercourse was but slightly restricted.

Squares of royal purple silk were brought in as seats upon the floor, but one or two ottomans came also, lest foreign guests should weary of the enforced position, — a thoughtful and by no means superfluous courtesy to some of the company, since sitting on the floor, although a desirable and graceful accomplishment, is supposably one not easily acquired. Personally, I enjoy it for unlimited hours.

Seated at length on the purple squares, with ottomans in reserve, stealthy shadows crept up to the bright room from the dusky garden, its paths and stone lanterns just visible in light from the ashes of sunset, while tiny cups of tea were removed, and the entertainment began.

Daintily made boxes containing sweetmeats were placed before all the guests, among them the Japanese and American flags shining forth in amicable proximity through clear yellow jelly. Examined and admired, they were set aside to be taken home at close of the evening; the first course of the dinner following at once, each of us had his individual table, unsplit chopsticks, and pretty waitress.

Japanese food is for the most part indescribable in English words; many of the twenty-seven articles which appeared during the evening were delicious and familiar Japanese dainties, and using chopsticks is a quickly learned and easily remembered art, but for certain choice and highly prized viands a severely acquired taste is requisite. In addition to soup and cooked fish, hot *sake*, raw fish with pungent sauce, chopped chicken, *daikon*, shell fish and chrysanthemum petals, lily-bulbs and rice rolled in rare seaweeds, there were also quail and French claret, lest, as our host observed, American appetites should suffer in the midst of Japanese plenty. Between courses the sliding paper screens shutting off the next apartment were withdrawn, and several choice plays were performed, the whole entertainment lasting from six o'clock until after eleven. These old classic plays are now kept up in Japan chiefly by actors who perform them for love, and their own satisfaction, as it is no longer the order of amusement which young Japan enjoys enough to assure financial success.

When the screens were first pushed apart, an archer was disclosed, handsomely dressed in the costume of old feudal days, — a haughty and impressive nobleman, engaged in stately conversation with his attendant, also in fine ancient dress. Very soon a man of lower caste entered, leading a monkey, and bowing low to the knight. After a few moments of dialogue the attendant told the newcomer it would be necessary for him to yield whatever his lord might ask, to which the man readily assented; having indeed no choice in those days when a nobleman's will was law.

The knight, fancying the monkey's skin, demanded it for a quiver to hold his arrows. But that request almost broke the man's heart, the monkey having been his nearest companion for years; he protested that he could not live without his little friend. A well acted scene followed in which the monkey's owner ventured humbly to remonstrate, telling the knight how they two had fared together, how he loved the little animal, and how hard it would be to kill him, although knowing he was bound by his promise to do so if the nobleman persisted. This was all so dramatically done that it hardly needed the clever running translation of our friends, — the story told itself in action; and when the man, looking tenderly at the monkey, told him he must die, that even he could not save him, the little creature — a small boy in reality — turned his head toward his master, looking up with unaltered confidence and love, and certainty that it must be all right if his dear master said so; it was a piece of acting so pathetic that the audience was greatly moved, and waited breathlessly for the end. Finally the knight's heart was touched, and he released the poor man from his promise, becoming so exhilarated with his own unexpected generosity that when the monkey in gratitude began to exhibit some of his choicest tricks, the lord was moved to vigorous imitation; and the play (called *kyogen*) closed with a series of cleverly performed feats of agility. Then the screens were once more closed, while farther courses of the dinner progressed.

The usual singing, and girls playing the samisen, went on at intervals, as well as songs by old men; and the famous *no* dance was superbly performed in the most elegant of ancient costumes.

Another sort of dance, in stately measure, called *gaisen*, followed, by three girls in black and gold, a celebration of certain victories in the late war. Afterward a comedy was acted between an old man in search of a wife, a "matrimonial agent," and a veiled female, who subsequently disclosed a hideous face. This play was called *fukitori*, or choosing a wife by playing the flute.

A famous juggler was next introduced, whose remarkable feats ended by producing handfuls of butterflies from nothing, until the whole room was full of the flutter of delicate wings; suddenly condensing, apparently, a magnificent white cock stood upon the magician's hand, and surveyed the company loftily.

Another play, later in the evening, related to incidents of the Chinese war, entertainingly interpreted by our faithful friends, and followed by the Maple Club dance, a graceful performance in which all the beautiful costumes were ornamented with designs of maple leaves,—as indeed everything is in the house itself.

The closing scene was charming; several pretty girls were scrubbing white linen, and beyond, a background of attractive landscape showed yards of similar linen drying. The whole thing finally resolved itself into a dance where all went through a variety of steps and evolutions together, flourishing the white cloth above their heads, twining and untwining the long strips in every variety of lithe posturing, with which the most ardent pupil of Delsarte could not compete.

All these performers, except the old men and the classic actors in the first piece, were girls belonging to the Maple Club. The charm of these professional entertainers, even in much simpler places than the Maple Club, is indescribable. But where everything is strictly high class, the maidens had an ineffable touch of dainty refinement. Late in the evening, dinner and entertainment having lasted over five hours, Chinese tea, water-ices, and lady-fingers closed the repast, in compliment to the foreign guests.

Riding away amid the pretty *sayonara* of the assembled establishment, our feeling of regret was most genuine that the stately, courteous, slowly moving life of the old days should ever give way before innovations of a busy modern civilization which all too soon will find no time for ancient customs. It is pleasant to see that the calm and unhurried politeness which causes acquaintances meeting on the street to stop and slowly bow low three times to each other has not yet wholly given place to the curt nods of the Western world, — all that the rushing life of an American business street seems willing to permit. Japanese men who still wear the graceful gray silk and black gauze native dress seem to preserve intact the spirit and expression of old time courtesy.

European costume at business or office seems to possess a curious power of imposing foreign manners therewith; although a long time must elapse before inborn graciousness will be sufficiently lost for a Japanese to be mistaken for a veritable foreigner.

The then American minister, Mr. Dun, Secretary of Legation during our former visit, was absent in America, but the Chargé d'Affaires, Mr. Herod, with his charming wife, omitted nothing in the way of Legation hospitality.

Their home, filled with treasures from artistic corners of Tokyo, undiscovered by the tourist; their white-robed native servants; windows wide open to the hot night; *punkah* wafting welcome breezes — how deliciously familiar and weighted with memory was the scene of that last dinner before the expedition departed for the mysterious north!

All necessary official arrangements made, — passports issued, apparatus safely stowed and started for Hakodate on the Sakura-maru, — the Astronomer, with the Musician, Chief, and their assistants, among them the second mate Andrew, the Japanese cook and his staff, also set forth in the same direction by train, with all lesser paraphernalia for science as well as enforced housekeeping in remote Kitami province. The photographer, Mr. Ogawa (also our photographer during the former Japan eclipse at Shirakawa in 1887), was to follow within a few days, and the interpreter detailed by Government would join the expedition at Sapporo.

Passes and official documents insured a more than obsequious attention from all railway employees, who speedily emptied an entire first-class carriage at Tokyo for the expedition, all of whom started off in the best of spirits for Aomori, the northern port of the main island. Thence a steamer conveyed them across the strait seventy miles to Hakodate, on the southern coast of Yezo, to meet the Sakura-maru, the members of the expedition joining their apparatus on board for the trip to Otaru on the west coast, where the special steamer Suruga-maru, already dispatched, took the entire party for the long voyage to Esashi.

CHAPTER XVII

IN FAMILIAR HAUNTS

<small>Nobody can revisit with absolute impuuity a place once loved and deserted.</small>

THE "unscientific contingent" of the Coronet party had classified themselves boldly; the expedition members proper were no less distinctly arrayed. One of the company, however, could not be absolutely identified with either. By no means learned enough to belong wholly to the specialists, her superficial attainments in the heavenly science prevented her unchallenged acceptance as one of the division which declared themselves "know-nothings." Her wise resolve in consequence, therefore, was to enjoy the best in the programmes of both. To this end she watched the expedition depart for northern Yezo in comfortable consciousness that after a month the eclipse camp would be in readiness to receive her and her humble assistance, — while happy journeys with the non-scientific friends would fill the intervening weeks.

Away from Yokohama changes in the last few years are not as apparent. To be sure, the genuine old swords, dispersed at first so carelessly, are no more to be picked up in every shop, and ancient robes of state and classic *kakemono* (scroll pictures) can be found no more at bargains; but the native life goes on much in its normal manner as soon as the immediate influence of the foreigner becomes less.

Primitive natives even now do not willingly eat three slices of pickled *daikon*, a favorite vegetable, since legend has it that a man doomed to death for some crime ate three slices at his last meal on earth. Two or four are therefore preferred. Watering the streets is still accomplished with much simplicity — by "joggling" out of a cart, scattering with dippers, spilling from buckets, or squirting with little force pumps.

The hotel where visitors formerly stayed in Tokyo was the Sei-yo-ken, near the foreign compound Tsukiji. It purported to be a foreign hotel, and so it was as to cuisine and beds; but it so recently had emerged from Japanese ways that it remained very picturesque, from the moist little entrance courtyard with ferns and growing decorations, to the Japanese attendants in native dress. Few now resort to the modest Sei-yo-ken, travelers being fonder of staying at the Imperial hotel, large, airy, and impressive, where the guest might imagine himself in one country as much as another, except for the jinrikisha in waiting near the entrances and their attendant coolies. If one now speaks of the Sei-yo-ken, he is generally supposed to mean its branch establishment, the attractive tea-house of the same name in Uyeno Park.

The park itself seems little changed; the trees are larger, the shady paths more beautiful; and even midsummer imagination can picture the wonderful arcades in cherry-blossom time — the whole air one exquisite haze of pink perfume.

In the great lotus pond buds were beginning to take shape amid rich green leaves; and the magnificent Golden Gate where ends a long avenue of stone lanterns was like a fresh creation, its superb restoration having been accomplished soon after our other visit.

Shiba temples did not fail of their earlier charm, where richly decorated altars show golden gleams of lotus and incense burner, vase and candlestick, with calm Buddhas gazing immovably into far-off space, and ceilings whose every panel is a separate study; with mystic odor of incense filling the dusky interior, and placid-faced priests moving silently about with shaved heads and ecclesiastical robes.

Formerly visitors had to remove their shoes before entering the cool, dimly lighted temples with their exquisitely lacquered floors. But the spirit of change has touched these immemorial shrines, and now coolies pull over the boots of foreign visitors a sort of soft white stocking, or *tabi*, which they tie deftly round the ankle, — a tribute, no doubt, to the constantly recurring visits of persons from over seas, to whom taking off shoes means a longer and more elaborate operation than slipping the simple wooden clog from the foot of a Japanese guest. Nine years before I had been an object of intense interest to a whole congregation from a temple service, and to several of the ministering priests as well, all of whom followed me out for the novel experience of watching a foreigner button her boots. But now the operation itself is no longer necessary.

The great tomb of the second Shogun, the largest specimen of gold lacquer extant, was to me even more impressive than before. More pathetic, too, were the memorials of the forty-seven faithful Ronins — more closely human and personal the swords, the worn old garments, *kakemono*, and ornaments; even their graves in the shade spoke touchingly of a loyal constancy ennobling to the annals of any country.

Certainly two visits should always be made to a distant land; the first grows luminous in the light of the second, and together they throw a clear radiance upon the inner spiritual meaning of its life and story.

That greatly abused word, picturesque, can perhaps be most properly applied in Tokyo to its moats, with banks sometimes turfed in vivid green, or walled with stone, from the top of which fine old pines lean their crooked branches low down toward the water. The white walls of the old Tokugawa Castle still rise in a tangle of gnarled pines, though the dwelling itself was burned in the Revolution (1868), and the streets make innumerable sudden turns through and around ancient fortifications, adding an immense charm to jinrikisha riding in the great capital.

In one of the pleasantest quarters stands the Peeresses' School, which, as its name indicates, daughters of nobles only may attend. One of its leading teachers is a brilliant young woman who, as a little child, was of that first famous group of Japanese girls (including the present Countess Oyama) sent to America for a foreign education more than twenty-five years ago. Living chiefly in Washington, where I remember her as a fascinating child of high-bred manner, studying later at Bryn Mawr, and returning to her native country while still young, she combines the best of both civilizations.

The little peeresses have a delightful spot for their educational efforts. They are charming girls, and the refined type of face is in striking contrast to that constantly seen thronging the streets. All wore native dress, often exceedingly rich and handsome, occasionally a royal purple *kimono* or *obi;* and their demeanor was exquisitely courteous and graceful, noticeably so even in a land where fine manners extend to all classes.

Modern culture and that of the old régime are here successfully united; and while thorough instruction in English was going forward in one room, in another a grave and elderly Japanese scholar was giving punctilious care to the intricacies of Japanese penmanship — would it were with us as much of a fine art! There a young girl was learning the elaborate form necessary in removing a *kakemono* properly from the wall; here, poetry and classics were studied faithfully by the gentle daughters of an ancient nobility.

All went on without fret or hurry; composed and gracious earnestness were everywhere apparent. At recess there was no noise, or shadow of confusion, but a great deal of bowing to teacher and guest as the classes filed out; and during their recreation the best of merry manners prevailed. The lovely garden was rich in verdure and artistic arrangement, and a soft rain gave additional freshness to every growing thing.

One well-remembered haunt (the Nakadori) for

picking up bits of old lacquer and bronze and porcelain, and ancient embroideries, seemed to have felt the touch of foreigners in more ways than one; and prices, still elastic, began at heights never dreamed a few years ago, descending with greater reluctance. Genuine articles, too, are rare in the little street, and modern imitations frequent. Still, with care and discrimination beautiful things may still be found in its precincts. Without sidewalks, each shop wide open, owners placidly smoking in the midst of their wares, the customer strolls along the way from one to the other, seating himself on the edge of any shop floor as fancy strikes him, his jinrikisha slowly following, its amiable coolie ready to assume entire charge of purchases.

Unfailingly reliable, no one of the multitude of wooden boxes containing vase or lacquer or whatnot, carefully tied with small twine handles for convenient carrying, is ever mislaid or unaccounted for by the *kurumaya*.

Of course in these attractive shops the purchaser must depend chiefly upon his own judgment of quality and value; but near by are two responsible places where prices are definitely fixed, each article precisely as represented. In one of them is shown the cloisonné without visible wires, invented by Namikawa, whose work is a dream of beauty. All in soft, delicate tints,

— dim moons with sprays of ethereal cherry blossoms (*sakura*) dashed across them, faint mountains against ineffable skies, with a suggested bird or two — the thought is in every case poetically conceived and executed, one large piece having occupied Namikawa for nearly four years. The old man himself, modest, retiring, and exceedingly refined, bears marks of the true artist in every expression and movement.

Across the street is a permanent exhibition of works of art by the leading masters of Japan in their specialties. Ivory carvings of wonderful beauty and skill, bronzes, lacquer, porcelains — everything is of the finest, with prices which may not be lowered.

A pleasant habit among reliable dealers in Japan when sending their bill to one person who has bought several articles of large value, is not to make a discount, but instead to give a "present," perhaps something admired by the customer but not finally included in his chosen purchases.

Only two or three days remained before the Coronet would start on her southern trip; one of these was the Fourth of July, a famous day in the happy port of Yokohama. All ships in harbor were lavishly decorated with countless flags and pennants, the American admiral gave a reception on the Olympia, noontime salutes of twenty-one guns from the men-of-war made the harbor reverberate, day fireworks filled the air with brightness, and a special tiffin at the Grand Hotel (where well-remembered, dusky Cingalese in tortoise-shell combs and flowing white draperies still displayed their wares on the veranda) was attended by scores of pleasant people. Early in the afternoon various feats of juggling took place on the lawn at the landward entrance of the hotel, and a baseball game was played on a fine field, between a Japanese nine and another made up chiefly of sailors from the war vessels. The Americans won, but the Japanese played well, their running, and sliding to bases, being particularly agile.

In the evening various events of a social nature went forward on shore and in harbor, brilliant fireworks and illuminations flashed over the quiet sea; and our national holiday became in retrospect dignified and invested with a certain elegance as well as crude patriotism.

But Kobe and the Inland Sea would not come to the Coronet, so turning her bow lightly toward the south, good-bys were said to the friendly harbor, and the pretty craft sailed airily off down the bay, and along the coast.

THE CORONET DRESSED FOR THE FOURTH OF JULY, YOKOHAMA HARBOR

U. S. S. Olympia

CHAPTER XVIII

SOUTHWARD

Praise the sea, but keep on land.
HERBERT, <i>Jacula Prudentum.</i></div>

IT was unmistakably typhoon season. The Coronet plunged at once into a gale, with higher seas than any experienced during our whole voyage across the Pacific. Even after seven or eight thousand miles of recent training, some of the company were unhappy because of violent pitching. A story was recalled at which we had scornfully laughed when first related on board : —

"Friends," said the captain of a steamer laboring in a fatal storm, "We must prepare for death. We shall go down in an hour."

"Heavens," groaned a passenger, "must we live an hour yet!"

After a tumultuous night breakfast-time found us back in Yeddo Bay, anchored farther down than before, near the light-ship and outside the breakwater. The morning sea was very calm and pale, and covered with small fishing-boats.

Our friends on the Olympia proceeded at once to engage in conversation by signal flags. Embarrassing questions were asked as to this humble return, after our refusal of several invitations on account of immediate necessity for reaching Kobe. All of which were answered from the Coronet with unabated cheerfulness.

The International Signal Code contains all one could possibly wish to say at sea; translation of remarks from the Olympia was quite exciting, as well as composition of replies, and selecting proper flags to express them.

In addition to a flag for every consonant, there is an "answering" pennant, one for "yes" and one for "no," with every sort of combination. "More help is required," for instance, is D C V B; "thanks," R S J; and sentences for all circumstances and conditions fill two or three hundred pages, with special appeals for help in emergencies, as "I am on fire," or "I am sinking."

For a few hours we lay still, until the wind had lessened outside; then, toward sunset, with B D R ("good-by") fluttering, the Coronet once more set forth, on a quiet sea.

Until darkness Fuji was magnificently in evidence, and constantly changing foregrounds made new pictures all the evening. Sometimes a steep, sharp bluff, then a line of soft green hills; once a large fleet of fishing boats seemed lying at his feet. Even the clouds in this artistic land are decorative, their long, wavy, golden lines like the conventional cloud-shapes of kakemono or embroidery or carving. Once a huge dragon in brilliant yellow lay just above Fuji's sombre crown; again, fleeing women, elusive mountains, and onrushing animals. For hours this splendid, shifting spectacle continued, about and above the grim, faultless peak, until night fell and land and water became one mass of quiet darkness under the starlit sky, a dull, volcanic glow from Vries Island touching the east with sullen light.

Almost a week passed in delightful but somewhat aimless sailing along the coast; baffling breezes alternated with dead calms, or what appeared to be beginnings of gales, with an uncertain feeling in the air, and typhoon color in the sunsets. Slowly the idea was accepted that yachting along the Japan coast in July and August affords unsatisfying recreation. Reluctantly it was abandoned. Although already far passed, a landing in Suruga Gulf was the most available place, at the little town of Shimidzu. Thither the Coronet's bow was repointed, that her owner and his guests might proceed overland to Kyoto and Kobe, the yacht afterward returning to her Yokohama anchorage to await our return.

But Shimidzu is a closed port, and whether this unexpected influx of foreigners could obtain permission to land was by no means sure. All one bright, sparkling forenoon the Coronet was beating up the beautiful gulf. Fuji splendidly dominates this whole region, and the bay is hemmed in by lesser hills and mountains, green to their cultivated summits, and touched by lovely haze. Thatched houses line the shore, and an occasional temple shows the fine lines of its roof higher in the sunny air, while terraces of vines and tea plants rise behind.

Word had apparently gone forth that a foreign vessel was coming, and from villages along the coast, fifteen or sixteen miles away, gathered the sampans — filled with a curious crowd, most of whom had never seen an American before, much less an American yacht. Closer they clustered, more numerous as Shimidzu was approached, gazing with undisguised amazement, entirely different from the spoiled sampan scullers of Yokohama.

Anchoring boldly in these forbidden waters, a native man-of-war was discovered near the shore, the red sun-rays from a central orb on the flag of the Imperial Navy fluttering gayly in the pleasant breeze. Very soon an imposing boat set forth from the vessel's side, and two officers came on board, one of whom spoke a few words of English. It was carefully explained to them through our interpreter, Okita, that baffling winds, stress of weather, in short, while on the way to Kobe, had necessitated our unexpected advent in their midst. Permission to land was asked, and at once and most graciously granted, even before our special passports were shown, with a great bundle of documents to Japanese dignitaries.

Omnipresent police also made their visit of inspection; but nothing could exceed the courtesy with which the yacht's company was treated, while scrutiny from thickly crowding sampans was entirely friendly, if still amazed. Officials in all departments of the government knew of the expedition; but to ignorant fishermen, and peasants surrounding us as we landed, we were an unexplained wonder, certainly novel and probably grotesque.

In a procession of seven jinrikisha, the little town was traversed, and we were out upon the Tokaido, toward Shizuoka, the nearest railway station, Alfred and Okita following to watch the two absurd tipcarts laden with our *kori* (baskets) and drawn by women. And Japan unadulterated and chiefly unadorned ran out to witness the passing. Young mothers with blackened teeth, and chubby babies on their backs, little sisters with heavy brothers on theirs, schoolboys in *kimono* well tucked up into their *obi*, and boys and girls without any *kimono* or *obi* at all; occasionally an old man arrayed in a garment of green mosquito netting — all flocked to the street as our train of *kuruma* went by.

It was a poor little village, yet the wide-open

houses were clean, and through the parted screens at the back could always be seen small and tastefully arranged gardens, dear to even the humblest Japanese. Making match-boxes appeared a prevailing industry, with silk spinning, and weaving cloth. Hollyhocks grew profusely, and countless blossoms of hydrangea were fastened on doorposts. Trumpet-vines flaunting great scarlet and yellow flowers covered many a little house — stately lotus was beginning to show fair pink buds in wayside ponds, and the shrilling of cicadæ filled the summer afternoon.

Rice-fields were full of cultivators, men and women and children, who straightened their bent backs for a moment, looking up stolidly at the passing jinrikisha, their dull faces hardly capable of expressing even surprise.

Toward Fuji the mountains were blue and hazy, though the king himself had withdrawn; and the road was lined with young cryptomerias, not a hundred feet high, like those bordering the glorious avenue toward Nikko; and groves of bamboo tossed their delicate green leaves in the warm air.

Shaded by ferns, and not very clean, — probably rioting ground for countless families of microbes, — streams of water flowed through the streets, beautiful if deadly; and over them leaned women, artlessly arrayed, washing vegetables in the running water. Occasionally some child or young girl would catch sight of this procession advancing far down the road; she would instantly vanish, rushing frantically for the rest of the family, only to return streetward in hot haste with grandmothers and babies to gaze till our disappearance. A few smiled amiably in answer to smiles of greeting from the jinrikisha, but many seemed too dazed to apprehend the fact of a common humanity.

Once a whole school passed, walking decorously, two by two, conducted by their teachers, the little girls in front in scarlet petticoats, the boys in gray divided skirts, with high, stiff belts. They examined the foreigners with interest, though devoid of rustic surprise.

Time in Shizuoka was not sufficient to visit its old castle, or temples. The little hotel has three or four "foreign" rooms, bare and unhomelike, the native portion neat and attractive like all good Japanese inns.

European food was served; but attempts to adopt and imitate things Western were pathetic; in the *tokonoma* (niche or recess), usually sacred to artistic *kakemono* and accompanying vase or bronze, hung a map of the Canadian Pacific railway.

The world is undoubtedly progressing, but in just which direction is not always apparent.

CHAPTER XIX

GIFU AND THE CORMORANT FISHING

> For flying at the brook, I saw not better sport these seven years' day.
> SHAKESPEARE, 2 *Henry VI.* ii. 1.

A NATIVE inn of especial charm is the Tamaiya at Gifu. Deliciously clean, the rooms open off shining corridors upon lovely outside verandas overhanging mossy garden courts, ponds full of goldfish, blue porcelain jars, stepping-stones, shrubbery, and stone lanterns.

The sliding screens of old gilt were decorated with spirited drawings of horses and scenery, beautiful metal ornaments, and fine carving. Ceilings were of delicate wood paneling, or paintings of flocks of ducks. Little closets or cupboards for the few dainty conveniences in each room, bedding, mosquito nets, and so forth, had doors decorated in monochrome drawings; the *hibachi* (braziers) were exceptionally handsome bronze. The *tokonoma* had each its fine scroll picture, and stand of lacquer holding an incense burner, or perhaps a porcelain vase of tall grasses or spray of blossoms arranged with the consummate skill of typical Japanese art. The proprietor, a man of much refinement, was a collector of ancient paintings.

On moonless nights from May to October cormorant fishing is in progress upon the Nagara river at Gifu, a spectacle very popular with Japanese travelers. After it was quite dark jinrikisha from the inn conveyed us through the city and across the river. The little shops were wide open, and many persons coming and going through the narrow streets. Flaring torches have largely given way to lamps, and an artistic paper lantern may sometimes, in these latter days, be found intimately associated with a modest incandescent light. But the *kurumaya* carry their tall, narrow lanterns, and run very rapidly.

Many queer little turns around dark corners brought us to a long bridge which by no means went straight across the river, but had several curves and angles in its passage over the Nagara. Small boats on the dark water beneath slowly drifted down stream, burning a few boughs to attract fish; but these were merely amateurs.

Below the bridge could be heard a rush of falls, and a dark and heavily wooded hill rose high against sky but a shade less black. A sharp turn brought our whole picturesque procession to a halt at the farther bank, where a large native crowd had collected, and our pleasure-boat lay awaiting us.

Everything was in holiday attire, and stepping on board the decorated craft we felt as if a natural part of this festive scene. Nearly all the boat, except a high and pointed bow, was taken up with a pretty, matted room under a light wood ceiling, the sides of paper screens now pushed widely apart, the opening draped with pale blue curtains and blue and pink lanterns swinging all around the roof.

Swarthy and half naked coolies immediately pushed off into the river, our boat becoming one of a fleet, the others filled with Japanese pleasure parties, also being poled up the stream. The river was very wide and dark; far across was a shingly beach; beyond, a high, dusky hill.

Three attractive *geisha*, engaged to entertain us before reaching the scene of cormorant fishing, now took gracefully upon themselves the duties of hostesses — tea first and a musical programme after.

The oldest played the *samisen* (three-stringed instrument); the second, who sang, was about sixteen, wearing an enormous and curiously tied *obi*, long enough to reach the floor, and many ornaments in her hair. The youngest could not have been over thirteen, but her hair was also burdened with scarlet and silver and golden adornments, and with her palm she beat a little red, tasseled drum. Both the younger girls were in crape *kimono* of blue and scarlet, and their names signified "the small wave," and "the sweet bell of a Shinto shrine."

Various songs and dialogues were performed, and simple but graceful fan dances. With bright scarfs tied over their hair, they assumed pompous expressions and went through one humorous little play. Many of the words were merely nonsense syllables, and the melody was easy to

 etc.

remember. The music is founded upon the harmonic minor scale, and melodies rarely end upon the tonic, which has apparently no musical value in Japan.

Suddenly out of the darkness boys and men appeared in startling nearness, walking by in the water, their bare brown legs glistening and their dark blue *kimono* tucked high up. The effect was curious, to say the least, — boats and walking figures close together in the same stream; but feeling a slight scrape, and looking over into the water, it was found very shallow, with a shadowy bed of variegated pebbles.

Singing frogs made lovely music all through

the merry evening, and as it grew later the little maids finally prepared a Japanese supper, — eels and rice, fish and seaweed.

After a while a certain commotion up stream indicated we were near the famous fishing. Six brilliant lights seemed drifting downward, and in a moment they surrounded us. Six boats had each an iron cage swung forward over the water, full of brightly burning wood which threw a wide glare. In each bow stood a man holding twelve cords attached to as many cormorants, large, black water birds, struggling and screaming and diving in every direction. Not to tangle all those lines required the skill of a circus driver, as each bird went its own way in search of the fish it instantly swallowed.

But the unusual part of this method of fishing is that a heavy iron ring at the base of the cormorant's neck is so tight as to allow only the smallest fish to pass through. All others lodge in the throat, and when that is full the bird is hauled back into the boat, and made to disgorge what it has just been at such trouble to obtain. That a bird should thereafter immediately desire to go fishing again seems odd, but its ardor is unabated, and it rushes once more into the fray with ever new enthusiasm. Three thousand of the *ai*, a sort of trout, is not a large evening's catch for a single boat.

The scene was unique, — flaring faggots, half naked boatmen, the dusky river full of brightly lighted pleasure craft and moving figures, baskets of shining fish, and the excited and fluttering birds.

Each man is greatly attached to his cormorants, and if by any chance they have not managed to swallow enough small fish for proper nourishment, others are given them from the catch for a good supper. Then they are tucked into basket cages to rest until the next night's sport.

And so pleasure-boats and fishing-boats drifted down the river together; the jinrikisha were waiting, and through dark and quiet streets, over the long bridge and around unexpected corners ran the little procession, dashing into the Tamaiya courtyard soon after midnight.

CHAPTER XX

KYOTO

The pine is the mother of legends.
                    LOWELL, *Reverie.*

"MADAM," said a courtly Japanese gentleman to an American single lady of uncertain age, "you remind me of our beautiful pine-tree."

"Ah!" she replied, visibly flattered, "and may I ask in what way?"

"Because, although you are so old, you are ever green," he answered suavely, quite unaware that he had failed to pay her a supreme compliment.

This incident came to mind when rolling comfortably through the city of Kyoto, across the rushing river, which seemed to have as much dry and stony bed as actual channel, and past innumerable temple gates toward the Yaami Hotel. Glorious conifers thickly covered the surrounding hills, and the hotel itself is set in a background of towering cryptomerias, sombre, stately, beautiful. Truly one might be compared to many things worse.

The famous cherry-tree in the city park was surrounded by an amiable, strolling crowd of women and children, and in a moment the outer gate of the Yaami was reached, and the upward walk, by mossy rocks, under large shade trees, up steps, past ponds and fountains and lanterns, led us to one of the verandas.

The hotel stands on varying levels, to which there are many approaches. One may traverse a piazza, and entering, ascend ordinary stairs; or by an outside stairway, and corridors overhanging a delightful little public road which looks like a forest path up to some mountain deity's innermost shrine; or he may walk farther through the garden, past another pond and up a few more mossy stone steps set with vague, artistic tiles, reaching thus the second or third story of another portion of the house. Tall evergreens clothing the hillside close by shelter temples and shrines enough to occupy many days without once visiting the city below.

From the upper verandas the view has a truly magnificent sweep, taking in all of Kyoto, usually wrapped in a bit of dreamy haze, and the far, green-blue hills beyond — at night mysterious and impressive with myriad twinkling lights, under a young moon sailing in the high heavens among lightly drifting cloud. Fireflies flitted through heavy shrubbery below the balconies, murmuring water tinkled softly in the warm darkness, and the humming, buzzing, singing of insects in the garden filled the summer nights.

Many foreigners were enjoying the artistic surroundings and excellent table of the Yaami — Americans, French, English, carrying national peculiarities as distinctively and carefully as if precious enough to pack in *kori*. Late every afternoon, while taking tea in the breeze of an upper balcony, in cool *kimono*, we watched various parties of indefatigable shoppers and sightseers toiling up the steps, their faces red and hot with exertion, but a familiar expression of satisfied bargaining upon their moist features, as if reflecting upon the purchases dutifully carried in armfuls of little wooden boxes by jinrikisha men following.

Radiantly beautiful were early mornings on the hillside. The exquisitely pathetic sweetness of a near-by temple bell often rang out at dewy dawn and over the silent city, with a call fit to beckon the soul straight out and away — anywhere, if only might be reached its realm of peace and forgotten pain. Its tone was a sacrament. Fragrance from freshened gardens rose to the balconies, and finally the sun, still partly shrouded in morning mist, shone on the glistening verdure, and mere living led everything to rejoice with exceeding gladness.

Okita was very entertaining; almost he might be reckoned as guide, philosopher, and friend. By profession an interpreter, he was invaluable in all situations. Even half the legends and histories he related about shrines, persons, and scenes would have made an amusing and entirely original volume upon the flowery kingdom. In temple, shop, or castle Okita was equally at home. Only at sea was he ever overcome by circumstances.

He was very comfortable upon the subject of religion, of which he said he had none, — Buddhist nor Shinto nor Christian. He laughed a great deal, with a funny little pucker of his nose quite irresistible. The Captain was "famous fellow" now, he said, both yacht and expedition had been written of so much in Japanese papers. Priests ministering at a Buddhist altar he called "sacred fellows." When asked about his family Okita laughed heartily. "Had a wife one time," he said, "very nice woman — very nice. But I too young, so divorced her, ha! ha!"

The Kamogawa is wide, shallow, sunny, crossed by countless bridges. Always there were children tumbling about in the water, women washing, and lines of houses close to the edge, their balconies overhanging the stream. And there are canals bordered by willow-trees, and moats, and ever mountain backgrounds and birds flying decoratively against yellow sunset skies.

Kyoto is credited with eight hundred temples. But it was Kyoto in large, impressionist effects —

shrines, palace, castle, arts, scenery, all blended in one glowing memory rather than clearness of minute detail — which time permitted in this royal city of the centuries. Built five hundred years ago, the golden pavilion, its ceiling, walls, floor, balcony, and rafters overlaid with precious metal, was the favorite haunt of the Shogun Yoshimitsu, whence he often gazed enraptured at the moon, or at' the opposite hills, once covered with white silk for the pleasure of an Ex-Mikado who wished to imagine snow in summer heat. A dreamy morning was spent here, and in the beautiful grounds, a funny little boy reciting in high, artificial singsong, with a sudden drop into normal tones at the end of each sentence, legends and history which must have lost much of their quaintness in translation, as he conscientiously pointed out the springs where Yoshimitsu bathed or drank and made tea, and every other especial spot.

Myoshingi Temple is lofty beyond others, with almost a cathedral effect of space; Nishi Hongwanji has a superb series of apartments once used by the daimios, and decorated with golden screens; Higashi Hongwanji was founded shortly after the other, about three hundred years ago, and burned just before the Revolution. The present edifice, only recently completed, is of noble proportions, and gorgeous decoration, —

abounding in panels of lotus on dull gold, and gleaming altars filled with rich art.

Worshipers came in constantly, clapping their hands to draw the attention of deity, a cere-· mony practiced with apparently equal effect upon any absentee, whether god or servant. The old Hawaiians also clapped their hands in praying, perhaps for the same reason, though their gods were of a less cultured variety. Buddhism hardly seems decadent when a new temple of such richness is built and maintained by modern enthusiasm.

San-ju-Sangendo, as its name implies, supposably contains 33,333 images òf Kwannon, goddess of mercy. Actually about a thousand, they represent the others by various computations, and as row after row of these golden ladies rises, one behind the other, each statue nearly life-size, the effect is overwhelming. An old attendant in the temple told us solemnly that all were miraculously made from a single willow-tree, pointing out also what he termed "devil-protectors" on either side of the central figure. Certainly they should be effective in warning off all the powers of evil.

Of course the goddess in her multitudinous representations has an occasional accident, and a hand or an arm must frequently be replaced. The divine repair shop was discovered, but not even by an offer of *sen* galore, nor by manifest appropriateness to his profession, could our good Doctor prevail upon the attendant worthy to present him with a discarded nose or even a stray finger.

Famous bells, among the largest in the world, fountains springing miraculously to avert conflagrations, historic carp, altars mystic with the incense of generations, — appreciation became almost pain, as day after day went by and we realized that year after year instead must pass before half could be seen, much less assimilated.

I wish that young, middle-class Japan, in transition state of costume, would not allow itself to appear incongrously projected upon a background of temple or castle walls. Practice with European clothes should go on remotely from these great monuments, "the finished fashionings from a far past."

From A. D. 793 to 1868 Kyoto was the capital city, and the buildings of the Mikado's palace cover many acres. But of far greater beauty is the former castle of the Shoguns; without, fortress-like and stern, even though the moat was in places filled with lotus in the glory of its blossoming; within, full of art and magnificence. In striking contrast to the plainness of the palace, it is a commentary upon the relation between Shoguns and nominal Emperor before the Revolution. Its rooms are a wilderness of golden screens painted by famous artists, with peacocks and pine-trees of natural size, tigers and strange birds. The celebrated "wet heron" panel is not in very good preservation, though still marvelously fine; two sparrows upon another are so natural that they once flew entirely away, Okita assured us, coming back of their own accord. Special permission must be obtained to enter either castle or palace, and guests inscribe their names in a book within the gates, where also may be read directions respecting behavior. Visitors are expected to "leave their overcoat, mitten, stick, walking-cane umbrella or whatever they take with them" to their own servants or the attendant before entering the buildings.

The shops and industries of Kyoto — who can describe or resist their fascinations! Delicious cups of tea welcome the purchaser, pretty *sayonara* attend his departure. Memorable are the rare and odd conceits of the vases and bowls of Seifu, the first ceramic artist in Kyoto, and descendant of famous potters; and cloisonné to rejoice the soul is made by Namikawa. Of the same name as the inventor of cloisonné without wires in Tokyo, the two are alike in earnest and poetic feeling, enthusiasm, and utter absorption in their art.

Namikawa's house is an education. In room after room of spotless neatness and beauty sit a few workmen on the floor, each with a tiny table holding wires, enamel, brushes, and all the paraphernalia of the art, sliding screens of glass opening upon a garden and pond, lovely if diminutive. Shrubs and flowers conceal an odd bamboo fence shutting out the city; the pond has miniature rocky cliffs on diversified shores; gold-fish and carp swim fearlessly as near the guest as possible; brilliant blossoms brighten a corner.

Upon our exclaiming over its beauty, the dear old artist said simply, "The workmen must have it to rest their eyes." Memory called up the scenery, with a few noteworthy exceptions, provided for resting the eyes of those employed in American industries,—piles of ashes in rear enclosures, varied by tin cans and an occasional old boot.

This prevailing love for the beautiful in all classes in Japan was well illustrated by the cook of an acquaintance in Kyoto, an illiterate man whom she one evening discovered sitting quietly, long after the hour, beside his untouched dinner. His reply to her question as to why he forgot his meal time was characteristic. As he pointed to the sky, with a radiant expression on his worn old face—"Who could eat," he exclaimed, "with such a sunset as that to look at!"

Truly we should entreat that apostles and missionaries of the beautiful be sent us from Japan.

Namikawa's cloisonné is worthy microscopic

study. His backgrounds are largely rich lapis-lazuli. Shapes and decoration have much variety, though few pieces are kept on hand, this famous work being largely ordered, or bought in advance of completion. Of the specimens finished, several showed white cranes, long clusters of conventionalized wistaria blossoms, or the popular iris. The whole process of making, too, was watched, from the first design sketched upon the copper vase, to the final, often fifth or sixth polishing of the repeatedly fired enamel.

The Nishimura embroideries were as fine in their way, — one particular screen remaining in memory as an almost perfect work of art. Its three panels, about six feet high, represented in solid stitches a thickly wooded hillside. The feeling for each sort of verdure was exquisitely portrayed, — the deep pines where each "needle" was shown, delicate maples, and lighter foliage, yet the whole effect broad and noble. Above the hill was a pale blue sky full of shreds of trailing mist, some of which had drifted down across the trees, — an effect constantly seen in Japan, — while in the right-hand panel a magnificent waterfall tumbled, white and foaming, from a height, flashing through the green to spread itself out in a tumultuous brook beneath, flowing off and away through the other panels of this masterpiece. No one could imagine, without seeing it, that embroidery could be so wet, or a cloud of stitches so filmy. In quite a different way, Nishimura's cut velvets are scarcely less beautiful. Here, design and coloring are woven into the fabric, whose threads inclose tiny copper wires. Finally a workman with a small and exceedingly sharp knife cuts carefully along the top of each wire, making actual velvet of portions to be rich and dark in effect, but leaving uncut distant Fuji, skies, pale moons, or shining water. The wires carefully withdrawn, a modern but most lovely work of art is produced.

Although vacation time, the Doshisha (One Purpose Company) was visited, that university founded by the late and greatly beloved Neesima, an Amherst graduate of 1870; and the girls' school close by, where a few pupils and teachers were found. The girls sang for us some weird native melodies, remarkable harmonies being supplied by a foreign teacher at a small organ. Harmonizing Japanese airs is an almost untried musical field, offering many curious opportunities for original effects.

In the eighth year of Meiji (1875) the Doshisha was opened at Mr. Neesima's home, with eight pupils. Through untold discouragements this "puritan of the Orient" struggled on with his beloved institution, only to leave it at his death in 1890 firmly established and prosperous, a tangible legacy from his devoted life, a monument to the pervasive power of his magnetic, unswerving personality. Through him Amherst College had become more widely known in Japan than perhaps any other institution, even before its later graduates, Kanda, Kabayama, Uchimura, Sawayama, and others had also carried its fame to their native land.

The Shinto festivals are full of beauty, despite the original simplicity of this faith. Happily timed was our Kyoto visit for one of these characteristic celebrations, a typical *matsuri*. The special day was July seventeenth, but all through the week the city wore a festive air, every house showing its new wooden bracket with a roof, under which hung a huge lantern. At twilight all were lighted, and gay drops of crimson or golden brilliance flamed as well in arches, festoons, high loops along the buildings, — glowing, pulsating, quivering strings of tamed and decorative fire in luminous figures. It was a fairy scene.

The evening of the sixteenth was particularly fine. Rockets flew hither and thither; countless globes of pale or scarlet flame in double rows lined every street, theatres were ablaze with brightness and gaudy pictures, sounds of music and drum coming from within to the happy, surging crowd. Many pictures were very amusing, — one showing a huge man engaged in throwing people over a precipice who, in their unwilling descent, took all sorts of queer oriental attitudes. Everybody was full of merriment, the babies out in full force. Little stands for shaved ice were popular centres, and varieties of disastrous, cooling drinks flowed freely.

Dwelling-houses, even the smallest, were in gala array, tiny interiors wide open to the street. Their usual straw mats were quite hidden, sometimes by rugs, more often by heavily woven cotton with white storks or ducks on a dull red ground. Exceedingly decorative, my attempts to purchase one always failed, their owners declaring them heirlooms; very old, exceedingly valuable, greatly prized, and only used during festivals.

The walls of every room were hidden by handsome gold or white folding screens, painted or embroidered, the *hibachi* being the only article which might be termed furniture in the room. So the whole effect was orderly and beautiful. In daytime the house fronts were decorated with floating curtains or strips with blue and white horizontal bars or other simple design.

The universality of interest in the festival, the personal eagerness and pleasure shown by all, were delightfully refreshing.

On the morning of the seventeenth an invitation came from the owner of a house on the route of the procession, to witness it from his roof. Upon reaching the ridgepole, sunshine was blind-

A "FLOAT" IN MATSURI PROCESSION AT KYOTO

ingly hot, the gray tiles scorching, but a little platform for two persons was shaded by a big paper umbrella as canopy, and a brisk breeze tempered the heat. The street below was thronged, the procession just having reached Narachu's house as we arrived.

High wagons or "floats" (*dashi*) draped with superb brocades and embroidered temple-hangings went slowly by, sometimes surmounted by a growing pine or cedar, or perhaps the life-size figure of a man in classic armor or other old costume, engaged in brandishing a branch of cherry blossoms in the face of an enemy upon imaginary battlefields — a " poetical fellow," explained Okita.

Under the canopy of the float men and boys beat rhythmically on small drums, singing and throwing tufts or branches of " good luck " to the crowd, in the shape of green leaves enclosing sacred rice-cakes. At the front of one float, three girls, thickly powdered, were performing some stately ceremony; on another two men stood on the projecting platform dancing fan dances.

Children and young girls taking part in the festivals are not allowed to carry a parasol from their houses to the rolling cars, even if it rains, and not infrequently their handsome costumes are quite ruined. Each street is responsible for a float drawn by coolies living in its precincts, accompanied by gentlemen on foot who also live in the street, — high-class worshipers at the temple, in their cool, gray silk ceremonial dress, carrying fans and wearing flat straw hats. At frequent halts their servants set down little stools upon which the gentlemen rested for a few minutes.

The huge, unwieldy wooden wheels have no means of being guided; so a coolie or two crouched along beneath like a new kind of coach dog, putting sticks under the wheels to turn them slightly to the right or left when they ran too near the happy crowd.

'T was a merry time, light-hearted as to inhabitants, sunny and fragrant as to weather, picturesque and characteristic as to processions and decorations. Fair Kyoto, with your long, long story, your immemorial temples, your gay religious festivals, your mountains and pines, your exquisite art, your gardens and river! Beautiful Kyoto, sayonara!

CHAPTER XXI

NARA

> Framed in the prodigality of nature.
> SHAKESPEARE, *Richard III.*, i. 2.

> Nara, the Imperial Capital,
> Blooms with prosperity,
> Even as the blossom blooms
> With rich color and sweet fragrance.
> JAPANESE POEM.

A HEAVY, tropical downpour had set in, with no cessation for days. Uji, its temples, famous tea-plantations, beautiful lotus pond, and Phœnix Hall (a reproduction of which was sent to the Exposition at Chicago) were seen through such a whirl of descending waters that, except a general impression of beauty, its memory is blurred and misty. Great danger of floods prevailed all over Japan, breaks in the railway line were constantly reported, and Nara was no exception to the general condition, looking half-drowned as we approached its historic groves.

Rich in temples and monuments, its sitting Buddha, fifty-three feet high, is larger than the one near Yokohama; the rich material is said to abound in gold and silver, yet as a work of art it seems less impressive than the Kamakura Daibutsu, partly, perhaps, because that is in a noble park, while the one at Nara is dwarfed by the ancient temple inclosing its massive proportions. An escaping thief is reported, by the voluble coolies about, to have lived safely for three years by sacrilegious retirement into its nose. On each side are figures described as Myo-i-rin kwannon, on the left, "watcher of the noise of the world," awaiting with calm patience one word of wisdom or eternal truth from the babel of humanity; and on the right, Kokuzo-bosatsu, god of the universe, holding one hand aloft.

Early in the eighth century Nara became the capital, and was thenceforth known as Heijo, or castle of tranquillity. No less so now, more than a thousand years after, "the noise of the world" scarcely comes near enough to its peaceful groves even to be watched. The little inn was charming. The rooms opened on the customary outside veranda, whose polished floor led to a few wide steps of green turf, — the entrance to a garden, somewhat larger than common, where quaintly arranged stepping-stones, bronze storks in various attitudes, and groups of pine and bamboo formed the foreground for a placid lake lying beyond. Still beyond was a thickly wooded shore, here and there a pagoda or temple-roof amid the trees, and mountains over which drifting cloud laid softly trailing fingers of mist far down their green sides, "as if one might climb into the heavenly region, earth being so intermixed with sky," as Hawthorne wrote long years before of another land.

Even the bronze storks looked wilted in the continual rain, — feathers bedraggled, attitudes dejected.

A noticeable feature of Nara is its tame deer, wandering in street and park and temple grounds. Secure from harm, these sacred animals walk confidingly up to any passing jinrikisha, in anticipation of a liberal meal, which the rider is supposed to purchase from women standing conveniently near.

The Kasuga temple grounds are shaded by enormous and aged cryptomerias, making an impressive archway. The sun had briefly emerged, and countless rushing brooks and cascades filled this lovely spot with a cool murmur of falling water. Mossgrown stone steps lead up to shrine after shrine, past myriads of stone lanterns placed as offerings by the devout, or in memory of friends, — perhaps to win heavenly favor for themselves. There are nearly two thousand of these lanterns, tradition relating that the oldest was given by Kobo Daishi, a famous priest and author of the Japanese alphabet; an expert as well in the fine art of penmanship. He constructed the alphabet in a poem, which roughly translated runs : —

"Even the colors of flowers decay, and this world is like a dream.
Nothing is constant, but we should not be asleep because the world is like a dream."

I-ro-ha, the Japanese word for alphabet, or syllabary, opens the poem.

We chanced on a special day at Kasuga Temple when a sacred dance was just being performed by young girls, whose ceremony was exactly and gracefully executed; dignified posturing and bowing, precisely in unison, alternating with complicated evolutions with fans, and sticks of little bells. Two or three priests sang, and played upon flutes, sometimes a perfect sixth above the voices; so there was actually a suggestion of harmony; though more often an interval utterly unmusical to Western ears.

The girls were heavily powdered, but close to their hair the thick whiteness ended abruptly in a curve sharply defined. Their eyebrows were shaved, painted ones high on the forehead giving a curious expression of wondering innocence; and their black hair, ornamented over the forehead with large white artificial flowers and twinkling pendants of gilt, was tied tightly back and allowed to fall straight down from the neck. They wore scarlet skirts and white *kimono* with figures in

STONE LANTERNS AND CRYPTOMERIAS AT NARA

gold and blue; two *eri*, inside kerchiefs of white and of scarlet, lay against their smooth throats. Very sober and dignified were these maidens, — as utterly in earnest as the solemn priestesses of the tea ceremony, with but a shadowy smile occasionally visible. This dance is called *kagura* (heavenly enjoyment). I paid one *yen* for my share in the lofty amusement, and a *yen's* worth is sufficient; but if one choose, he may stay all day and spend fifty *yen*.

Four gods are worshiped at Kasuga Temple, ancient personages in Japanese mythology with memory-defying names, — Takemikatsu chi-no-mi-koto, Futsunushi-no-mikoto, Amatsukoyane-no-mi-koto; while the fourth is a goddess, Himiōngami.

For over seventy years Nara remained the capital of Japan, before Kyoto, but the Mikado's local palace has long since disappeared. Partly burned, the remainder was bought and carried off bodily by merchants and carpenters, for use in building their own houses. Altogether, seven emperors lived in Nara, and Kasuga Temple, built by Shotoku, stands near the sacred hill Mykasa-yama, often sung in Japanese poetry.

A fine pagoda, nearly as large but not so richly carved as the one at Nikko, belongs to still another temple, the Kobukuji; it was built by the Empress Komyo, over a thousand years ago; and everybody tells you that a superb pine before it is at least as old.

Nara possesses also a very sacred though rather restless and impatient white horse, with bright blue eyes and bushy mane, an albino among animals, securely fastened in a very small shrine, tail where one would expect the manger. But it is easy for the passing traveler to feed him with beans for a quarter of a cent, and to worship or not as he chooses. This was but the fifth horse encountered in two weeks, so we were not forced to grieve actively over one of the articles in our passports, which forbids attending fires on horseback. Japan is a land of fires, but not of saddlehorses.

Struggling masses of turtle, goldfish, and carp contend for a bit of bread thrown into the pond (Sarusawa no Ike); a sheet of water chiefly famous because centuries ago a certain young girl at the palace, thinking she had lost the Emperor's affection, here drowned herself. The monarch, coming afterward to its banks, composed in her honor a poem, now cut upon a stone slab standing some distance out from shore; its waters, he recites, can never become dry, because composed of the maiden's tears.

Another story told by Nara people partakes more of the supernatural. A certain governor called Shijo caught one of the sacred tortoises from this pond, thinking to bake it in his pan, foreseeing an especially delicious meal. But when

TEMPLE AT NARA

he took off the cover — behold! the imprisoned dainty had taken itself off miraculously, saved by compassion of the god of Kasuga.

Summer rain descended softly during most of our days at Nara, — increasing occasionally to heaviest tropical intensity, and on the way to Kobe the whole country was practically under water. Rice-fields and gardens were submerged, little houses made. islands of themselves, and small boys, arrayed with simplicity impossible to excel, paddled about ecstatically among trees and over fences. We afterward found that the entire island was seriously flooded, and railway travel everywhere interrupted.

CHAPTER XXII

YACHTING IN THE INLAND SEA

A level floor of amethyst,
Crowned by a golden dome of mist.
LONGFELLOW.

SINCE it had been found impracticable to take the Coronet into the Inland Sea and she had returned to Yokohama, the Captain had chartered a native steamer, the Miyako-maru, for the Sea trip, a craft somewhat larger than the Coronet, and carrying a crew of twenty-eight.

Built exclusively for native use, the fittings of the Miyako-maru, including staterooms, height of ceilings, and galley appointments, were diminutive in scale; causing much merriment in the company, two of whom occupied the saloon, where they could stand upright if directly under the skylight. Ordinarily that apartment would carry thirty or forty native passengers. But the vessel was entirely new, and satisfactory, even if, as in Japanese inns, lesser toilet arrangements were in the public eye, and confined to one small brass basin, a pitcher, and a tumbler.

Members of the crew were greatly interested in the foreigners, which class they had heretofore had slender opportunity to observe, — nor did they neglect the occasion.

The unclassified member of the party had at last to face a genuine division of the ways. To be sure of seeing the eclipse, now two weeks away, she must start at once upon the long journey to Esashi, more than fifteen hundred miles intervening between Kobe, where the Miyako-maru awaited its passengers, and expedition headquarters in Kitami Province. But starting at once meant abandonment of the trip through the far-famed Inland Sea. Methods of possible travel to northern Yezo were an unknown quantity; how many days would be used in getting there entirely uncertain. Fear of being late for the eclipse prevailed, and reluctant good-bys were therefore said to the non-astronomical friends (who still hoped to reach Esashi by the ninth of August) at the very entrance of their trip through Japan's enchanted waters.

Three of the voyagers upon the Miyako-maru have kindly lent me their Inland Sea journals in describing certain places not usually visited by foreigners, and a combination of portions of the three records has been effected, with occasional verbatim quotations.

During the first day Awaji was passed, which according to tradition was earliest formed of all the lovely island group in the Sea; and on reaching a land-locked harbor on the south shore of Shodoshima, the Miyako-maru cast anchor off Nomamura. Near by, and easily reached by a short row, is a tiny island with a *torii* and a neglected shrine to the goddess Benten.

The scenery grew constantly more beautiful as hours and days went on, islands clustering so thickly that "exit seemed impossible, and entrance a dream," strait after strait opening and closing in vistas of loveliness. Inland Sea currents were found very swift in places, twisting unwieldy junks around like toy boats. Even the staunch Miyako-maru was occasionally forced to hug the shore.

At Tadotsu much curiosity was evinced as to the strange visitors; and a train was taken there for Kotohera, to visit the Shinto temple of Kompira on a green hillside, where a god especially presiding over the fortunes of seamen is worshiped with unusual zeal. The Miyako-maru's native captain repaired thither at once, paying his devotions, like most sailors, with much fervor. Five hundred and seventy-two stone steps lead to the temple, — a warm climb on a July day. All the way little shops offered trinkets for pilgrims, and sacred horses demanded tribute from the faithful. The fine view and breeze rewarded the travelers, resting at the summit under grand old trees. The temple is simple; one of its buildings contains a green *gohei*, the paper prayer of Shinto, and a mirror; there are, also, many paintings of scenes in storms, and fanciful accidents from which this deity is supposed to rescue his faithful worshipers.

On a point just beyond Tomo a little temple to Kwannon was charmingly situated, approached by a covered stairway. Onomichi, "Tail city," is stretched along the narrow channel farther on, with several fine temples on the hillside, — a labyrinth of islands, atmosphere dreamy, colors exquisite.

Ondo *seto*, or strait, could not be passed until full tide, which gave opportunity to anchor in a little bay where fishermen were spreading a net across the current, singing as they drew it. Passing the narrow opening at any time seemed an impossible achievement. The "hidden door" is exceedingly narrow, the current swift. On one side a stone lantern stood out in the water; on the other a village so near that a pebble might have been tossed into its street.

Wandering at one's own sweet will through the beauties of the Inland Sea, unrestricted by traditions of regular trips taken by average tourists, is undoubtedly pleasant, save an occasional drawback. The native captain suggested anchoring one night in the harbor of Kure; during the late war, and still, an important naval station. Restrictions against foreigners are so severe that special passports would have given the Coronet no permission to enter this harbor; but as the Miyako-maru was a Japanese vessel, sailing through these halcyon waters under the full sun flag, no trouble was anticipated. The journals tell the story: —

"As we were preparing to anchor, a launch from the station came out, an officer boarding and demanding of the captain what he meant by coming in without showing special signals, giving his name and other information. It seems that all vessels, even in passing the harbor, must show these flags, and our captain had committed a grave mistake, as our vessel was so new he had not yet received them from the Admiralty. He was informed that in this case he had no right to come in at all — and his reasons were asked. He replied that the vessel was chartered by foreigners who were traveling slowly through the Inland Sea, and after much parley as to our purpose and destination, the launch returned to shore for instructions from the head officer. It soon came back, with a subordinate officer on board who marched to our bridge and took command, ordering us back through Ondo Strait. Passing the narrow channel once more in safety, he remarked that our captain was apparently experienced enough in these waters to have known better than to make the awkward blunder of entering Kure harbor without his flags.

"We cast anchor just where we had been an hour or two earlier, and then heard that our captain must be taken back to appear before the authorities at the station.

"It looked seriously as if we might be delayed a long time, as the captain was liable to a fine of at least $75.00 and withdrawal of his license — even possible imprisonment; and we were all practically under arrest, and might not be able to finish our cruise in time to get to the eclipse station in Yezo.

"Expostulation with the officer now began. The objects of our pleasure trip were detailed, and what a serious matter it would be to delay us was shown; our special passports were exhibited, also letters from the governor of Kobe asking especial courtesy from governors of all these provinces.

"This explanation happily worked a good effect. In consideration of the papers showing what 'famous fellows' we were, he said he would let the captain off. The young officer was very polite all through, assuring us that the difficulty did not concern us in the least except in delaying us. Okita added to possible horrors of the situation by telling us that the captain's children would probably have been given away for adoption, his wife divorced, and the home broken up. Instead he was graciously pardoned.

"We steamed as far as Nakashima in beautiful moonlight, thinking it wiser to get away before a possible change of mind at the naval station. But we went on the outer side of this island instead of braving the dangers of Ondo Strait for a third time. We all sat on the bridge and sang, under a brilliant night sky, gliding through water which sparkled like gold lacquer."

A most beautiful as well as celebrated spot in the sea is the sacred island of Miyajima. Approach to it was in the early morning, when a rosy mist of sunrise lay between the steamer and the hillside. Close to the shore lay boats of fishermen, who sang one refrain while in the boat drawing the net, and another when pulling it up on the beach.

The island is unmistakable from its unique feature, the famous *torii* of camphor wood, which at high tide stands well out in the water. And it is shrouded in an atmosphere of more legend and romance than hovers about either of the other famous places, which with it comprise the *san-kei*, "three great sights of Japan."

At flood-tide the temple seems to float on a silver sea; and all the little dwellers of the deep pitch their tiny tents on its stone piers. Bridges and galleries connect the shrines and different buildings, and boats can be rowed up to the very holy of holies. When the tide is out, stepping-stones enable the pilgrim to go through the courts in more prosaic fashion.

Built sometime in the sixth century, the temple passed from Shinto to Buddhist and later to Shinto again, and fires and fanaticism wrought sad havoc. The relics are now carefully preserved and watched; and many beautiful things have been added in later days. Far aside from the line of that tourist travel which has despoiled many another spot once full of the poetry of old Japan, Miyajima bids fair to revel in legendary atmosphere for long years to come, — with sacred deer, no less tame and half-human than the pretty creatures at Nara, with innocent-hearted priests, and the reposeful silence of leafy maple groves.

It was just before the annual *matsuri*, which here is celebrated upon the water, with boats instead of decorated cars as elsewhere. Preparation for this great occasion was in active progress when the mystic galleries of Miyajima were visited. When, on festival night, eight hundred lamps of the temple are lighted, and masses of people assemble with songs and rejoicing, it is a resplendent scene, depending upon high tide for point of departure.

In this enchanted island no one dies, and no one is born. Sadness and pain are ferried across to the Aki shore. Blissful serenity has been its portion for centuries.

Okita, ever faithful to the proprieties, although personally lacking any marked religious preference, as he amiably announced, threw a glittering ten *sen* piece into the shrine, just in advance of two somewhat aged worshipers, who, clapping their hands to attract the attention of the god, muttered some unintelligible prayer probably for good luck. Apparently they thought the deity had smiled upon them rather quickly, for depositing one *rin* at the sacred spot, they quietly removed Okita's bright money, making by this transaction ninety-nine *rin*. As they were calmly departing, Okita, notified of the fraud put upon the gods, called after the ancient couple energetically that they had taken consecrated money — whereupon returning, they smilingly threw it back.

Shimonoseki is full of historic interest, from the time, in the third century, when the Empress Jingō started from Toyoura near by to conquer Korea, until the twelfth century and the battle of Dannoura; more recently still, the bombardment of Maida by the allied fleets in 1863, and most lately of all the stirring scenes of the war with China. Fine forts, guarding the harbor in both directions, fleets of junks with sails spread, and the channel shut in by steep hills bristling with black guns, offer a sharp contrast with the sunny silence and peaceful enchantment of Miyajima.

The Miyako-maru approached the city, stretching three miles along the narrow margin of land between sea and mountain, through a strait which was a scene of much activity during the Chinese war, transport ships all starting there. In the Fujino tea-house opposite, a delicious luncheon was provided, beneath the large room in which the treaty with China was signed, April 17th, 1895. Li Hung Chang occupied a temple close by, and in the street outside he was shot while being carried in his *kago* from the tea-house to the temple. His boys ran on with him to the steps, and alighting, he remarked, as blood trickled down his face, that he doubted if ever before a foreign ambassador had been assassinated while negotiating a treaty. Count Ito, and Count Mutsu, formerly minister at Washington, stayed at the Daikichi, an inn on the principal street, below the tea-house.

At evening the voyagers asked for an upper room as a "moon-gazing" place, thus gaining a fine sight of the full moon rising over the hills across the strait. Shimonoseki is picturesque at night, with paper lanterns swinging in the breeze, but not overclean or fragrant.

Clothing in these regions was scanty, — occasionally a woman was happily taking her bath in a tub set in the middle of a street. One's modesty seems able to survive seeing people with slight raiment, or with almost none at all, but when in addition, as sometimes happens, they shave their heads, it becomes positively shocking.

A place even less conventional was Beppu, on the return voyage to Kobe. Kiushiu Island is famous for hot alkali baths, supposed to cure leprosy and other ills. Here men, women, and children were partaking of this benefit indiscriminately, in the public tanks, while in a large building sat others waiting, their clothing left neatly in boxes along the wall. From the deck of the Myako-maru people could be seen bathing on the beach, digging holes for themselves in the sand, or sitting in the warm water with umbrellas over their heads.

A police officer, sent on board to make sure that all was right, seemed rather confused at sight of foreigners, and being shown the passports was manifestly unable to determine what to do with them. He confessed frankly at length that he had never seen one before. On shore Americans were equally strange, and, as in all remote Japanese towns, troops of people, young and old, followed in a lively procession.

The intention was to remain anchored off Beppu until midnight, that Matsuyama might be reached early in the morning. But at evening, the sailors having had leave during the day, which was Sunday, the Miyako-maru became the scene of various incidents. The journals again tell the story: —

"Most of the under officers and crew had been drinking *sake* on shore. Being in port, neither captain nor first officer seemed to have proper control. We attempted to keep back one of the men who tried to come aft without his clothes; one of his friends took his part; sampans alongside were selling more *sake;* and as the crew gradually came on board, girls from the tea-houses escorting them, shouts and hilarity from forward grew apace. Lest the entertainment should wax riotous we decided to weigh anchor and get off at once, thinking it less safe to remain than to trust the navigation of a drunken crew.

"Accordingly the whistle was blown and the siren given, but it was not until eleven o'clock that all were on board, and we could start. We took turns on the watch all night, — some on the bridge, others at the engine room; while the Doctor slept across the entrance of the saloon as a guard to the ladies: a sort of 'devil protector.' Sunday rest was not found beneficial to the crew.

"When we awakened at five o'clock the men were at work as usual, and everything apparently quiet. We abandoned Matsuyama, on the island of Shikoku, being afraid to give the crew leave again. Instead we went straight on to the whirlpool between Shikoku and Awaji.

"Anchoring off Tubi, a sampan took us to see the rushing current of Naruto Channel, — less a sight than anticipated, probably because the tide was setting in the wrong direction. Landing after a hard pull, we scrambled over boulders like the New England coast, and up a steep hill, where a fine view was met, of islands, strait, and far blue sea. After this an ideal cruise back to Kobe, where home messages again annihilated space and time."

CHAPTER XXIII

EXPEDITION EXPERIENCES

> O, what a load
> Of care and toil,
> By lying use bestowed,
> From his shoulders falls who sees
> The true astronomy,
> The period of peace.
>
> EMERSON, *The Celestial Love.*
>
> Ah! well I mind the Calendar;
> Faithful through a thousand years.
>
> EMERSON, *May-Day.*

THE northern voyagers had made no sign for many days, except an occasional telegram as to progress in the novel journey. But just as I was starting for Esashi, and the travelers in the south were about to embark upon the Inland Sea, and the experiences related in the last chapter, a journal arrived, in which Chief had minutely chronicled, for our edification, an account of the daily adventures of these scientific gentlemen.

Beginning faithfully with their departure by train from Tokyo in the heat of that first day of July, the outline of their story follows, in the veracious words of their historian:—

... "The cook was in a second-class car and the mechanic in a third-class car, and our grub in the baggage car. Stops were so short and our command of the language so limited that to get either the cook or the mechanic out in time to tell the baggage-master that we wanted to get out a package seemed a very serious undertaking, and several stations were passed without anything accomplished in the commissary department. About four o'clock a man came along with packages of Japanese luncheon, consisting of two neat wooden boxes, one containing cooked rice, the other a variety of other food, such as *daikon*, ginger root, a kind of omelet, seaweed, — which looked like fine-cut tobacco and tasted as though the same had been soaked in fish oil; also a kind of dark brown substance of the consistency of jujube paste, but of quite a different flavor. We invested in some of this, — but there was plenty left. Then we all became thirsty. There was a small table in the middle of the car supplied with a pot of water, and three tumblers. We were afraid to drink, and here your devoted servant distinguished himself by volunteering to get beer.

"At the next station he found quart bottles that looked as if they contained beer, and he understood the girl to say they contained beer; so he bought them and returned to the car triumphantly. Upon opening the first bottle, however, it was not beer, but *sake*. We mixed some of this with the water and drank, but with sad counte-

VIEW ON THE RAILWAY NEAR MORIOKA.

nances. It naturally followed that the others 'had fun' with Chief.

... "Soon after this a determined attempt was made to get at the package of eatables. No one could remember the size or shape of it, so it was necessary to get into the baggage car and make a thorough search. Andrew had the checks. At the next station I hunted up the cook, and the mechanic appeared from somewhere; by the time we got the baggage master to understand the situation it was time to start again. Finally by locking Andrew up with the baggage master from one station to the next we found it.

"About this time they lugged out the little table containing water, and replaced it by one containing an outfit for tea.

"We made a nice evening meal with crackers, potted quail, tea, and so on.

"There was room enough in the car for us partially to stretch out for our night's rest, and sleep came sooner or later. I was some time getting into the land of dreams, and it required some miles to take all of me away from Yokohama harbor and the Coronet.

"The next day was cool and comfortable, and we arrived in good shape on time at Aomori. There was plenty of irksome duty here, finding carts and sampans to get our traps from the station to the steamer for crossing the strait to

Hakodate. We all had to act as vanguards or rearguards to see that nothing was lost.

"On the steamer, finding that no food of any kind could be obtained, we decided to return to the town and take supper at the tea-house. We were able to get omelet, boiled eggs, chicken cut into small pieces and cooked with onions, the latter being very good, except they had put sugar in it. As you were not here to give me a game of chess, I partook freely of everything.

"We had a very merry time here. Everybody tried to speak the language, and the girls in waiting were inclined to be sociable. When Andrew and one of them conversed, one in Russian, the other in Japanese, it was very amusing. . . . We returned to the steamer about 9.30, sailing at ten for Hakodate, and arriving there at five in the morning.

"At Hakodate we found our special steamer had not yet arrived, so we landed everything. Fortunately the hotel was near the pier, and there was not much trouble. About nine o'clock the steamer arrived, and Mr. Thompson came on shore. [He had gone with the apparatus from Yokohama all the way by water.] They had had rough weather. In the night the packages got adrift, and one of them struck Mr. Thompson on the head, making a slight wound. He is all right now. As *pro tempore* doctor of the expedition I examined him, and so report. . . .

"Otaru, July 4th. In the afternoon I went with the Professor by rail to Sapporo. . . . We went to a large hotel on the European plan, and were delighted to find delicious strawberries and fine cherries. Before dinner we called on the governor of Hokkaido. . . . Sapporo, July 5th. Soon after breakfast the governor arrived with Mr. Nozawa, who has since been detailed to accompany us and remain a few days at our station. Everything we expected was accomplished. The governor will write to the local governor at Esashi to receive us and assist to the best of his ability. Soon after the governor left, Professor Nitobe called on Professor Todd. He is connected with the Imperial Agricultural College at Sapporo, and he married an American lady from Philadelphia. Professor Todd returned with him. Otaru, July 6th. Left for Otaru at 9.35 A. M. in company with Mr. Nozawa, above-mentioned. He is to remain with us a few days, and afterward make a tour of inspection through certain portions of Hokkaido in the interests of fisheries and oyster beds. Also I had with me the student Mr. Oshima, and a police official, as permanent guard at Esashi. . . . We sailed about 2.30. The captain and officers are agreeable, and do everything for our comfort. It is very cool up in this region — too cool for comfort, in fact. There is much talk about the flies and mosquitoes we are expected to encounter in camp.

The prospect of being enveloped in a veil of netting hanging from the rim of one's hat, and having the face anointed with a mixture of castor oil and tar, is not inviting. . . . Some work is being done on board. A heavy wooden frame for counterbalancing the three-story instrument platforms is in process of construction, and parts of the pipe connection are being screwed together to save time at the station.

"At sea, July 7th. . . . A strong wind blowing and the sea coming up. About 8 A. M. we ran into a place called Wakkanai, to telegraph to Esashi as to sea and weather at that port, as there is no harbor at that place, and it would be impossible to unload our traps with the present conditions. . . . We are to wait in this locality until there is a change of weather. I don't like it. Professor Todd takes it calmly, however, and we are doing pretty good work on board. I have donned my overalls and jacket and help a little.

"July 8th. . . . The weather moderating toward night we got under way with the intention of feeling our way to Cape Soya, and anchoring just inside the cape if too rough to venture outside.

"July 9th. Just a solid month before the eclipse. We did not go outside last night, the wind having increased somewhat; about ten o'clock this morning, however, we started for Esashi. It was rough work rounding the Cape Horn of

Japan. . . . Esashi, July 10th. This has been an eventful day, inasmuch as we have finally reached Esashi, taken possession of our camp, have everything unloaded and under cover. Professor Todd and Mr. Nozawa went on shore early in the morning, met the local governor, and arranged everything at short order. . . . The town itself is not very large, a fishing village, one or two Japanese hotels, a few shops. There is a very strong odor of fish, but our place has it less than elsewhere. There are small flies about, but I have n't heard any complaints from members of the party, and neither netting, castor oil, nor tar has been mentioned as yet. However, it is still cool, and the wind is from the sea. . . . The cook has such a display of hams, bacon, etc., in his quarters that it looks like a corner grocery. . . . The Commandant of the Alger and the French astronomers have called, also the governor of this province.

"Saturday, July 11th. The day has been consumed in getting up the piers for the main station, setting up tents, and opening crates that contain the portable house. The weather has cleared up nicely, and the sun was out at eclipse time this afternoon. . . .

"Sunday, July 12th. Just four weeks before the eclipse. It is a clear day, warm in the sun, but cool in the shade. . . . Tell 'Doc' that I came near having a serious case in my capacity as assistant surgeon. A day or two ago one of the party tumbled over a pile of tent poles, and came down. He did n't get up at once, and said his leg was out of joint at the knee. Instantly after he said 'It's all right, it has slipped back into place.' I was much bothered when it first happened, — I knew something should be done at once, but whether to have him pulled out straight or doubled up I was n't sure. As he was already doubled up I think the first would have been proper. When he said 'all right' I promptly produced the Pond's Extract and recommended rest. . . .

"July 13th. This morning I came to the front again rather unexpectedly. I had started work on those everlasting plate-holders again, when Professor Todd called out that my professional services as doctor were requested at the French camp. One of the sailors was ill, the Alger had gone off for a few days, and they had no surgeon. So I took my bottles and paper of instructions that 'Doc' provided, and went up there with Professor Todd, and the assistant who came down for me. I explained that I was not really possessed of a medical education, but they were welcome to the medicine and the directions for use. . . . While there they wished me also to look at a sick sheep. They have a number of sheep in a tent. I felt the sheep's pulse, but doubt if I got hold of the right leg. I recommended rest. . . . This afternoon the report comes that both parties are about the same. I'm thankful they're no worse.

"July 14th. We have two flag-poles erected, one for the stars and stripes, with Amherst colors, and the other for the Japanese flag. They are symmetrically placed at the ends of our inclosure. . . . I hear from Professor Todd that at a meeting of the good citizens of this place, it was voted that on eclipse day there should be no wood fires made. Either cooking will be done the day before, or charcoal used in place of wood. This is to secure a clear atmosphere. Work on the portable house goes slowly. Theoretically it should be put together in a few hours. Practically it takes a good while to get things to fit. . . . I find myself feeling a little depressed to-night. The cook gave us some Japanese soup for supper. Perhaps it's that. . . .

"July 16th. The portable house is about finished outside. The different tubes for the lenses are being made ready to be bolted on to the platform, and lots of small work — overhauling and adjusting the plate mechanisms — is going on. . . . We had some washing done. Of course we don't mind such a little thing as undershirts starched and trousers creased the wrong way. . . . My duties to-day have been verily like that of Jack-at-all-trades. I have taken up electrical business, connecting galvanic batteries. Then I play carpenter, and screw small boxes on to a wheel; then I paint a lot of square pieces of wood; and from that I go to cutting out rectangular pieces of black velvet, and gluing them on to the inside of the boxes. . . . As to affairs out in town — there seems to be a great scarcity of small change. It is impossible to get even a *yen* changed. To make a small purchase at the shop near here, I had to leave the *yen* and take a due-bill for the balance, to be traded out afterward. . . . The others still run me a little about mess affairs. At the table when anything appears they say, 'What's this coming, Chief?' As I have n't the least idea what it is, I say, 'A little surprise for you to-day.' When I *do* say anything to the cook there seems to be a misunderstanding. Seeing onions for sale in town I suggested that we have some occasionally. The very next night, as a last course, when we usually have canned fruit or preserves, he served up two stewed onions to each of us. They were very nice, but why did n't they come earlier in the meal? . . .

"July 29th. There has been quite a little excitement in our town to-day. A few days ago the village officer or mayor went to Mombetsu to get the Emperor's portrait. It has been presented to the village school. A new schoolhouse is to be dedicated on the 11th of August, and the picture is then to be displayed. Now it seems that when

the Emperor's portrait travels about, it must be treated with the same respect as himself would be. So this afternoon there has been a little ceremony connected with the landing of the portrait from the steamer.

"A new sampan, having a canopy draped about with purple, and roofed with white bunting, was towed out to the steamer, by another sampan pulled by a large number of men. Plenty of flags displayed, of course, on both sampans, and also many flags and red and white lanterns shown along the streets. The portrait was inclosed in a square box, covered with white cloth and furnished with four legs; and two poles were fastened to it, so it could be carried on the shoulders of two men. All along the route from the landing to the schoolhouse, little hills of sand had been previously placed. Just before the procession started, these were made into a path, so that the Emperor would have had new soil to walk on had he not been his picture. The box was carried by men in white *kimono* and black hats shaped something like a bishop's mitre. The school children with their holiday clothes and unusually clean faces looked quite sweet. They were marched down to the landing and formed into two lines, the girls on one side, and the boys on the other. As the portrait passed, the entire school chanted slowly the Japanese national anthem. Afterward they re-formed, and followed it to the schoolhouse. I could not avoid the impression that they were going to bury it somewhere. . . .

"July 31st. We are looking for Mrs. Todd daily now. . . . Professor Todd told me yesterday that he thought everything was going on well, and that all he had planned would be finished in time.

"August 1st. A steamer arrived from Otaru, but Mrs. Todd was not on board, neither did it bring us letters.

"August 2d. This would have been a good eclipse day. Advantage was taken of the sun's presence to run the glycerine clock. Professor Todd is very much pleased with its action.

"August 3d. Only five more working days before the day that must bring us the Corona or bitter disappointment. To-day has been fine for the most part. At eclipse time the sun was out in good shape.

"August 5th. Mrs. Todd arrived this morning, and we were all glad to see her. We are very busy, but hopeful. There is a chance to send letters, so I have caught a few minutes' time to close this last installment of my journal and send it. I haven't corrected the proof, so make out the best you can."

CHAPTER XXIV

THE TIDAL WAVE

<blockquote>
Ruin itself stands still for lack of work,
And Desolation keeps unbroken Sabbath.
Left by one tide and cancelled by the next.
               JAMES MONTGOMERY, *The Pelican Island.*

Yet must thou hear a voice, — Restore the dead!
Earth shall reclaim her precious things from thee : —
    Restore the dead, thou sea!
        FELICIA HEMANS, *The Treasures of the Deep.*
</blockquote>

WHILE the expedition was thus setting up its apparatus, writing journals for comrades in the main island, and preparing for the eclipse at Esashi — while the Inland Sea party was still exploring the remote bays and straits of those fairy waters, and studying the native character under new conditions, I was hastening northward from one to the other.

In looking about for a guide or interpreter to accompany me to the far wilds of Yezo, I had been fortunate in meeting a young Japanese, formerly a student at the Doshisha in Kyoto, who speaks excellent English, and is a good French and German scholar as well. He particularly loves astronomy, has used the telescope at the University, and is a member of a well known

German Society. As for astronomical books in English, he has read much; Newcomb, Chambers, Ball, Miss Clerke, and their brotherhood — all are equally familiar to him.

Modern methods of observing an eclipse he had longed to see, as well as the phenomenon itself. Although his social position as student and teacher is far above that of interpreter, he was willing to go in that capacity, even running the risk of temporary caste misinterpretation, for the sake of seeing the astronomers and their work at Esashi. His father was once chief of the island of Shikoku in the Inland Sea, and the boy's whole life has been spent in arduous study.

Difficulties in learning scholarly Japanese alone are very great, even to a native; and it is said to require no less than seven years for a child to become sufficiently familiar with Chinese characters to use them easily. Besides the purely Japanese alphabet, invented by Kobo Daishi, it is necessary to know about twenty thousand characters before the classics can be intelligently read, even newspapers making use of twelve thousand.

These must be memorized, and the eye and hand trained to distinguish and delineate the faintest curve or variation in a line. With their own literature rich in fiction, fable or mythology, legend, and poetry, it was no wonder that all this and a good knowledge of other languages and literatures filtered through it, should have made Murakami-san's cheeks pale and thin, his physical vitality largely burned out by over-exercise of brain. But he admitted no fatigue of any kind, and started joyfully on the long journey to interpret as might be required.

Very quiet and retiring, he preferred Japanese food on steamer and train, staying quite by himself except when needed, — practically little for many days; for a few Japanese words go far, most officials speak English in varying degrees, and travel is comparatively easy for unaccompanied foreigners.

The floods which had set all the rice-fields afloat around Nara and Osaka were widely extending, and I had finally to abandon my pleasant scheme of following exactly the route of the expedition, passing through the familiar city of Shirakawa, whose old castle was our happy seven weeks' abiding place many years before. The fine mountain scenery farther north must also be unvisited, for the railroad was impassable at certain points, and might require several days for entire repair.

So another Yusen Kaisha steamer, Tairen-maru, was taken for Hakodate, and possibly Otaru. How Esashi could be reached from there was misty but enticing, as I rather hoped it might be necessary to travel a few days by packhorse over hitherto untrodden wilds, a few Yezo bears in the background and the "hairy Ainu" as hosts. However that might be, the next immediate stage of the journey was clearly defined.

All along the eastern coast of the main island, as the Tairen-maru kept her steady way northward, were sad reminders of the tidal wave, now more than a month in the past. Supplies and money still being sent to the survivors, we stopped for two or three hours at Oginohama, near the southern end of the afflicted region, to leave a variety of necessities for the suffering people.

The town was dirty and sordid, but blossoming white lilies and purple hydrangea brightened it, and our own familiar clematis climbed all over shrubs and even trees on the mountain path leading past a pathetic little burial place. Just beyond was a Shinto shrine, full of *gohei* (sticks with fluttering paper prayers), a good many spirited drawings of a cock and hens, and a spherical bell with a thick red cord attached by which to ring attention from the presiding, but perhaps otherwise occupied, deity. A steep climb, accompanied by numerous little girls with babies on their backs, brought me to a larger Shinto temple with a mirror and rough drawings of horses. Overhead, tall cryptomerias shaded a spot doubtless charming in a sunny day, but rather too moist for comfort under gently falling summer rain.

Oginohama received no damage from the great deluge, its harbor being on the inner side of a long promontory. The havoc was greatest in small but open bays near by, where the water heaped itself to appalling heights.

Japanese papers and magazines were still full of pathetic details of the great catastrophe, theories for its cause and reports of assistance to the survivors. Inability to read the complicated characters describing all these interesting matters as we passed along the afflicted shore was an exasperating drawback, for an extensive current literature pertaining to this subject covered the table in the Tairen-maru cabin. Murakami kindly read the articles to me, which probably lost much of their graphic character in verbal translation. But the harrowing and realistic illustrations by native artists needed no interpreter.

The day of the tragedy, 15th June, — according to the Old Calendar fifth day of fifth month — was an annual festival; and in many villages the primitive seaside folk had been hilariously celebrating when singular noises were heard, preceding the melancholy interruption. Curiously enough, barometers gave no advance indication of impending disaster; but on the morning of the fifteenth, an old woman noticed that the water in her well had almost disappeared. She is said to have told her neighbors that a great tidal wave was coming, though no one paid serious attention to her prediction.

When the wave was actually advancing, three of them, in fact, running shoreward from southeast to northwest, the receding water is reported to have laid the sands bare for a distance of eighteen hundred feet, white and glistening gruesomely in the murky night. Wave length from the first monster to the crest of the following one was not less than from twelve to sixteen hundred feet. Ten minutes completed the entire devastation.

In Kamaishi the director of the telegraph office saw his entire family washed away before his eyes; nevertheless, safe himself, he at once proceeded to hunt for his broken and scattered instruments among the débris. Owing to his faithful bravery and presence of mind, communication with the outside world was soon opened.

The avalanche of waters swept three times into the town, the first most terrible. In less than two minutes all houses standing in the lower part of the town were quite swept away, and thousands of persons suffocated or battered to death. But three storehouses ("go-downs") were left standing. Had the approach of this fatal watery mountain been anticipated for even a few minutes, many who perished might easily have saved themselves.

The "chief officer" or head man of the town was conversing with three callers when they heard the roar of unfamiliar waters. Jumping directly out from the upper story he and one of his friends took flight for high ground and escaped, while the other two, waiting to go down by the stairs, were caught by the flood. Four steamers anchored near the shore were carried inland and stranded in fields, almost without injury. Schooners and junks in rice fields were a common sight all along the coast. One small boat was caught in the forked limb of a tree. The water was reported in many places as eighty feet higher than the highest tide ever known, while one village remained in complete and apparently permanent oblivion beneath the sea. A few persons saved themselves by breaking through the roofs of their houses and there clinging until washed ashore, hours or even days afterward. While those overwhelmed were chiefly poor fishermen who lost not only houses and household goods, but their small gardens and crops and all their nets and boats, many farmers also were ruined, and cultivators of silkworms; the food supplies of whole provinces disappeared. Papers were full of incidents, horribly tragic, and details of saddest meaning. One young girl in trying to save both her grandmother and a little child lost her own life and probably theirs as well; when her body was found one hand was grasping remnants of her grandmother's dress, while the other still held tightly fragments of the baby's little *kimono*. The prevailing and unalterable love and respect in which older persons are held in Japan is never more tenderly illustrated than in scenes like these. A native picture represents a man for an instant undecided whether he shall save parents, or wife and children. Characteristically seizing his aged mother, he is shown rushing with her in his arms to a place of safety, while wife and child, trying vainly to follow, are clutching at his dress in despair.

In one village the force of the wave was so great that one hundred pine-trees over ten feet in circumference were entirely torn away, leaving only their broken roots. But in other places men and women washed into the tops of trees were safely stranded — climbing down to the ground when danger was past.

More than one hundred and fifty victims were cast ashore upon Sankwan Island, and subsequently rescued. About the same number of convicts, released from jail at Okachi when the wave broke over the town, returned a few days after to the Miyagi jail of their own accord.

Work among the dead and dying was heroically carried on, despite conditions of great discomfort and even danger to the rescuers. The victims

THE GREAT TIDAL WAVE AS PORTRAYED IN A NATIVE MAGAZINE

under rains and hot June sun became almost at once unrecognizable, and owing to the prevalence of Shinto faith among the relatives of the dead, cremation could not be resorted to without doing violence to their feelings and principles.

Among so many tragedies the finding of a fish, a gold ring in its mouth still encircling a human finger, was mentioned simply — without comment. As a relief to the prevailing gloom, an account was given of a young woman in a hot bath when the wave reached her, being lifted bodily, tub and all, and floated to a place of safety.

Without rice, sent by the government at once and in large quantities, the survivors must have starved. Terrible bodily injuries, too, resulted in loss of life, through lack of physicians and nurses with medicines and instruments.

A small fleet of a hundred and sixty fishermen were at sea, off a village, the great waves passing harmlessly beneath; and they had no knowledge of the horrors overtaking friends at home until their return late at night to the awful scene of death. Another party of fishermen, equally unconscious, picked up, with much surprise, a floating child — then to their amazement another — and two or three more; at length one of the men rescued his own little son; tragedy hovered in the air.

Through hundreds of miles of devastation,

corpses covered the beaches, and others were continually washed ashore.

In Hongo the entire hamlet of one hundred and fifty houses was obliterated, the sole survivors a party of men playing *go* in a temple on a hill, and eight children, carried by the waters to an elevated spot and deposited. Later, others were found, thrown upon the opposite coast.

A passing traveler, putting up at an inn, was seized by four women in the watery rush, who clung to him so desperately that he was powerless to move. Oddly enough, this proved the salvation of them all,—the combined mass defied the power of water, and ultimately found itself on dry land. The survivors, hurt, dazed, half-wild, wandered up and down for days in tattered garments, like demented ghosts.

Professor Kochibe's theory of the cause of the calamity is probably the most scholarly of all. For some distance out from the coast the water is shallow, but it suddenly drops to a great depth. The cavity is called Tuscarora Hollow, and is no less than four thousand fathoms deep. The probability is that a large piece of this wall, or great cliff, fell off, detached by a submarine earthquake, thus causing the huge rollers. Two deep-sea cowries found ordinarily at depths of several hundred fathoms were discovered after the catastrophe far up on the shore, at the edge of the wave-line, one of them but just dead.

The first recorded damage by an earthquake wave is that of May, A. D. 869. On the title-page of a Japanese almanac in the 9th year Kenkin (1168 A. D.) is a representation of an earthquake insect, on its back a map of Japan, an oblong body covered with scales, ten feet like a spider, and a dragon's head.

After leaving Oginohama the Tairen-maru passed a gruesome relic, — a corpse floating by, bleached beyond possible recognition, but unmistakably once a living person. Later in the day two or three others floated by.

As we passed the lovely coast scenery, finally reaching Shiriya Light, and Tsugaru Strait, separating the main island from Yezo, it was unspeakable relief to depart from a region so haunting in its calamity.

The steamer carried many Japanese passengers, one of them an officer high in the Imperial Navy, members of the Board of Education, the editor of a popular magazine, and several dainty little ladies who kept mostly to their staterooms. The naval officer, especially interested in oceanic meteorology, was the only Japanese who cared to come to the European table. He and I with the genial Scotch captain of the steamer were often the sole diners, — a slight roughness reducing the number at once to these three. Captain Kimotsuki spoke little English, but seemed much impressed that a lady could be a good sailor, even that she could think of traveling to Esashi at all. He was himself bound thither to witness the eclipse; but most of the other passengers would stop at Hakodate or Otaru.

Much interesting talk — and the Japanese are often great talkers — went on in the cabin between the various scientific and literary gentlemen, upon matters of current importance, and it was most trying to get only now and then a suggestion of its drift from an occasional familiar word.

I asked Murakami to give me a few lessons in reading printed characters. In a day or two, dates on Japanese newspapers became intelligible; but while philosophical and intensely absorbing, it is a discouraging accomplishment. A lifetime might be spent in its acquirement.

Read from right to left, the first two characters stand for *Mei-ji*, the present era of Japan, which began with the Revolution of 1868.

日 八 十 月 八 年 九 十 二 治 明

(AUGUST 18, 1896)

Tsugaru Strait was passed, the Yezo shores loomed in sight; Hakodate Head rose majestically above the sea, its base washed by a snow-white fringe of surf, its mighty cliffs green from sea to sky, plunging their heads into the softly drifting clouds over twelve hundred feet above. Hakodate seemed to have wide and fairly clean streets, the houses built evidently with reference to more severe winters than prevail in the main island. Stones, nowhere used for paving, were chiefly lying on the roofs. A pagoda, a few barren foreign buildings, and the graceful lines of a temple roof broke the general monotony.

In the light of a fair northern sunset the Tairen-maru cast anchor, and was surrounded at once by the usual fleet of sampans; and with colors dipping to her from numerous Japanese steamers lying near, she proceeded to settle herself for a twenty-four hours' stay in the beautiful harbor.

CHAPTER XXV

IN PURSUIT OF A SHADOW

<small>What shadows we are and what shadows we pursue.
BURKE.</small>

WHO could watch that superb Hakodate Head lying close by and not desire to see the widespread view from its summit! Early dawn found me well up its side.

The path might have been along some New England hillside. Red and white clover grew luxuriantly, humble heal-all and wild geraniums, spiræa and serpentaria, and hydrangeas, blue and white. Maidenhair ferns and royal *osmunda* flourished side by side with yellow lilies, dear old cinnamon pinks of traditional grandmothers' gardens, and fair purple iris. Young oaks shaded bluebells, homely yarrow sent forth its pungent odor, and wild sunflowers gleamed at every turn in the path. A large bush, full of splendid scarlet berries, was an unfamiliar member of the elder family.

On a mountain spur the view emerged from shifting fog in sunny brightness; just below us two reservoirs, farther away the bay, with a long, narrow beach separating it from the ocean, scores of vessels at anchor, and the gray town climbing a short distance up the hillside. Surf was beating high on the sea side, fog drifting off to south and east, glints of sunlight turning the water here and there to silver. Tall cryptomerias, huge ledges, a fine park and museum, a farmhouse where a pretty Japanese girl dispensed that rare luxury, rich milk — are all blended in one memory of Hakodate.

But not an Ainu had appeared.

The Tairen-maru, bound for Otaru on the west coast of Yezo, finally steamed off, past Shirakamisaki Point, and another stage of the journey was begun.

At Otaru (sandy road) women carry the burdens, and mushroom hats prevail. It is the port of Sapporo, whence a short line of railroad leads to the site of the Imperial Agricultural College. An attempt to reach the province of Kitami overland from here would have been seriously difficult if not practically impossible, the island consisting of roadless mountains, unexplored forests, and bridgeless rivers. Interesting as such an unusual trip might have proved, the time would hardly admit dallying with the unknown after such a fashion, and my first anxiety in Otaru was to inquire whether a Yusen Kaisha steamer might not be ready to make the infrequent trip along the northwest coast, around Cape Soya, and to villages along the northeast shore of Yezo. First information was disheartening — a steamer had some time before started for this remote region and the one now in port was intended for another direction. But it was found that a little steamer belonging to a local native line was liable to start for the north in a few days, if storm or rain or fog or high seas did not prevent.

And this seemed the only practical way to reach Esashi in season.

While this somewhat uncertain prospect was probably producing a thoughtful cast of countenance, as I revolved the possibilities of being late for the eclipse, a swift messenger arrived from the Yusen Kaisha office, — bringing the compliments of the manager, and the intelligence that he had received word from the head office in Yokohama to detail for me a special steamer for that far northern voyage. It would sail within two days, passes would be sent, and European food provided on board. Meantime Sapporo could be visited, and perhaps some Ainu seen; so my prospect immediately began to lighten.

Can it really be that Japan is seen from the windows of the train bound for Sapporo? To be sure the cliffs and hills on one side and the sea on the other were beautiful, fishermen's gardens held brilliant hollyhocks, and portulaca carpeted the ground. Marshes had cat-tails, and the green hills trailing mist over them. A rice-field or two disputed the landscape with Americanized fields where haying was in progress. Solemn crows perched upon gables; flagpoles and dead trees all had their brooding black occupants. Yellow toadflax lined the track, and evening primroses, large and golden, droopingly awaited sunset.

But Japan — where had that poetic country gone? Perhaps only itself under warmer skies, brisk northern breezes may blow away its most elusive charm. At all events, Sapporo seemed the most American city of the Empire — as indeed it well might, since the Imperial Agricultural College was established in 1876, under the direction, as organizer and president, of the late Colonel Clark of the Massachusetts Agricultural College. The model farm with its buildings, and the whole atmosphere of the place are American, not Japanese, with an effect practical rather than picturesque.

Buildings erected in Japan under the title "foreign" are apt to be bare and barrack-like structures, quite lacking the architectural grace which our older towns show, yet having lost the attractiveness of Japanese houses. Many such are in Sapporo.

Jinrikisha are few, and horses prevail on the wide unpaved streets, bordered by low buildings and small streams of running water.

Built for the Imperial Household in foreign style, but under Japanese management, the Hohei-kwan is a large hotel facing a typical garden most "restful" to the eyes. The cook is an artist, the attendants delightful. Here I found myself already famous from connection with the expedition.

In the usual native inns pretty maids conduct male guests to the bath, a custom as old as the Odyssey; male servants taking charge of the ablutions of ladies. The same custom prevailed at the Hohei-kwan, an amiable old man escorting me to the bathing apartment with much simplicity; he saw that the tub or tank was filled with actually boiling water, showed all the arrangements for comfort and convenience, and only stopped short of offering to take my *kimono* while I stepped in. That experience was reserved for the Etchuya Inn at Otaru.

Professor Nitobe and Professor Miyabe of the College, both of whom were educated in America, called at once, and through their kindness many interesting places were intelligently seen, among them the Museum, full of Yezo material, and — as everywhere throughout the Empire — triumphant mementos of the Chinese war.

The mysterious Ainu nation is exciting much attention in Japanese ethnological circles, and utensils, ornaments, and clothing are now being properly collected, preserved, and classified.

I, too, began to gather similar articles for Professor Morse of Salem, who had asked me to collect them for the Peabody Museum. He had hoped that I might obtain a primitive loom, bows and arrows, elm-fibre garments, musical instruments, and other Ainu articles.

Through Professor Nitobe's skill and courtesy I was enabled to purchase a number of rare relics, which induced great elevation of spirits, since Professor Morse had mentioned casually when asking me to collect Ainu material, that they especially dislike parting with their possessions, and some articles would be almost impossible to obtain. Here, however, I already had the nucleus of a fair collection, although I relied chiefly for the best additions upon the Kitami Ainu, where a foreigner would be a novelty.

Professor Miyabe, one of Japan's finest botanists, cleared up my doubts upon various shrubs and flowers, puzzling in their differences from and resemblances to American species; and altogether the time in Sapporo was far too short.

An event in one's lifetime is the first sight of an Ainu. A "civilized" specimen soon crossed my path, the most extraordinary figure in my experience. With his wild head of electrified black hair parted in the middle and standing out under a round-crowned and very dingy Derby, huge hoops of brass or German silver in his ears, his face largely hidden by an enormous beard and mustache, a white cotton *kimono* and cowhide boots, this anomalous relic of a vanishing nation was infinitely more pathetic than his veriest savage kinsman. His son, evidently a cross between Ainu and Japanese, — a peculiarly barbarous combination, — wore regular schoolboy "gear." Speaking in pure Ainu to his son, who answered in a sort of mixed dialect, the father was a melancholy and out-of-place specimen.

Sad it is to see a whole race disappear, — overpowered peacefully and half-unconsciously by a stronger nation of brighter intellect; but this is inevitable in the world's progress.

Bearskins in Sapporo were tempting, thick brown and golden yellow; but the hunters have to go farther for them now than a few years ago — more deeply into the northern forest.

In Sapporo, too, it was that I had a sudden awakening as to using my native tongue. As a little shop was passed, at whose open sides hung pretty Japanese brushes of many dainty kinds, I remembered my hearths and open fireplaces in remote Massachusetts, and exclaiming: "Oh, those brushes are dear! I must have

AINU COUPLE, THE WOMAN WEARING CEREMONIAL BEADS

one or two," approached the smiling and bowing shopman.

"Oh, no," said Murakami, gravely following, "You will not find them dear. They are very inexpensive!"—which was indeed the case. But the unintentional lesson was no less pungent.

A delightful Philadelphia gentleman of the old school, and his daughter, were at the Ho-hei-kwan for the summer, having lived for several winters in Tokyo. An invitation to their table, with many other kindnesses, made the hours homelike and gracious.

On the train returning to Otaru, Murakami-san gave another mild shock to his companion, with his superior use of English, at all times a model. Wishing to test my own impressions of the reason, I asked him why all the carefully built fences at the stations were invariably burned — pickets, posts, all charred some distance up from the ground. His reply was characteristic:—

"Carbon is not soluble in water," he said quietly, without farther explanation. But it was sufficient.

The Etchuya Inn is a fascinating little hostelry, its maids dainty, its native food of the best, its attendants more than attentive, its "head steward" unremitting in politeness. To be sure its bath-water was red-hot,—actually bubbling; the screens surrounding the tank had a row of glass panes in the middle, and screens have no locks. But its hand basins were of artistic, shining brass with decorative characters in the bottom, its tiny brass mouth bowls unique; and a long, polished corridor where water was superficially used by all the guests in common, opened to a green and blossoming garden court.

Chasing an eclipse, and then chasing an eclipse expedition, I deemed it appropriate to travel in a distinctly native way; so everything not absolutely necessary had been left on the Coronet, and my few belongings were packed in the pretty baskets, or *kori*. I had added, also, on the way, one or two white trunks of native manufacture; and from the Etchuya Inn, while seated of course on the floor, my letters were written upon long strips of Japanese paper with a camel's-hair brush and native ink. One epistle to America measured two yards and a half.

Meals, too, were served by smiling maids upon their knees, the fire-pot constantly replenished with glowing coals of charcoal, *o cha* (hot tea) always ready, and at night with the floor for bed, beneath Japanese green mosquito netting, I slept the sleep of Nippon, occasionally partly aroused by the pattering feet of mice or "twenty days' rats," as the Japanese call them. Lessons, too, were given me in correct trying of the *obi*, and old treasures of lacquer shown, among them the toilet set of a court lady of generations ago.

All the men and maids gathered with the chief steward at the entrance, giving farewell bows and fans, as we departed for the Kwankomaru. Murakami was silent, but apparently happy.

CHAPTER XXVI

STILL PURSUING

Shadow owes its birth to light.
GAY.

WHEN at creation a certain god and goddess were selected to evolve the island of Yezo from chaos, they were endowed equally with materials and ability to complete the task.

To the god were allotted the eastern and southern parts of the island, while the goddess was to attend to the western portion. They began together, vying amiably during the progress of their work. But alas, after the manner of women, the goddess one day met a female friend and stopped to chat with her. This friend, sister of Aioina Kamui (one of the most ancient forefathers, indeed the Adam of the Ainu race), must have been a seductive conversationist, for the two talked long and idly about acquaintances and neighbors, while the god at the east kept steadily at work, ever the custom of men. Looking up suddenly and seeing how nearly completed his portion was, and frightened at the state of her own unfinished regions, the goddess hastily threw together her remaining materials in a careless and slovenly manner, leaving this western coast in its present rugged and dangerous condition. But, add the Ainu in telling this legend, no one, even if disposed to grumble at the dangers of these shores, should presume to blame the Creator for such a state of things, as it is wholly the fault of his deputy and her tendency to gossip; and their lords often point the moral at women who talk too much, — " Set a watch over your lips and attend to your duties, for see how rough the west coast of Yezo is, and that all because of a chattering goddess."

The chattering, nevertheless, may be held responsible for a picturesque bit of work. Steep cliffs, often richly wooded with ancient trees, sometimes rise in bare and rocky impressiveness many hundred feet above the sea. Innumerable streams rush in white torrents down these majestic heights, using every ravine for their swift descent until the whole face of the coast appears laced with flying spray of continual cascades. Tiny fishing villages find precarious foothold at the base of cliffs entirely inaccessible, on beaches almost too narrow for the single row of thatched dwellings, even huddled against the steep rock behind; while constant surf, beating white and high before them, seems to make a village highway by the sea equally improbable.

Leaving Otaru seemed also leaving all fog and cloud. The Sea of Japan stretched clear and gray to the horizon, where a narrow strip of greenish-blue sky showed beneath horizontal lines of cloud. Fleets of fishing-boats lay in the offing, and toward the north hopeful sunshine, with an autumnal suggestion in its quiet beauty.

At Mashika a landing was made in late afternoon. Only a fishing village, it had lately grown to over five hundred houses and nearly three thousand inhabitants, such promise of financial prosperity follows in the herring's train. Mashika's "fire tower" was quite imposing,— a tall ladder rising high above the roofs, with a bell at the top, suggesting observation and alarm.

An official of the Yusen Kaisha came on board with a polite invitation to visit the town; and after a call at his house, with its beautiful inclosed garden where he made scientific tea, an opportunity was afforded for seeing a comparatively new Japanese colony. No foreign woman had ever been in the town before, and a tour of investigation about the streets and to the temple aroused an almost startling degree of interest in the younger inhabitants. The procession became more imposing in numbers at every corner. Hoping to escape from our following, we decided to visit an Ainu house, and, turning quietly off the main thoroughfare, as twilight was coming on, passed a rushing stream and took a footpath through deep bushes to the dwelling. But not so easily were the young people deprived of their foreign amusement, and every individual followed. Jumping the stream with alacrity, chasing single file through the narrow pathway, and actually arriving before us, they made a dense circle around; while the old Ainu, gray-haired and venerable, came out politely to speak to his singular guest. I counted sixty in the group, not including stragglers on the outskirts. An old woman was washing a big dish in the stream, — a piece of cleanliness learned with difficulty from the Japanese, since Ainu, away from civilized neighbors, wash neither themselves nor their clothes, nor utensils.

The old man spoke fairly good Japanese, and his story was a sad one, — told to his unusual and unexpected callers with a modest dignity. His father, so the tale ran, once lived in a very fine house, almost a palace to an Ainu, but it was burned, and his own as well, with all his treasures, so that now he was forced to live in the poor one where we found him. His oldest son had broken his leg, and all the father's money went to the Japanese hospital, while now his own eyesight was nearly gone — truly a pitiable plight for an old, white-haired Ainu.

When sunset had faded, and the landing or

hatoba was reached, — so near nightfall that the body-guard had considerably diminished, — this old Ainu was found waiting by the boat. Bowing low, he expressed in very good Japanese his sense of the honor done him by our call, and his gratitude and appreciation that so much trouble should have been taken by one coming from so far.

The men of the Ainu race are much better in appearance than the women, immense heads of bushy hair parted in the middle, and great beards imparting an impressiveness far from unpleasant. The women appear stolid and indifferent.

Our sampan lay in the surf, and a single plank, dancing up and down on the waves, connected it with the shore. A few lanterns gleamed here and there as coolies ran about, and bidding the old Ainu sayonara, the sampan pushed off on the dark water to the brightly lighted Kwankomaru lying at anchor outside.

A most beautiful feature of the voyage to Soya, northwestern cape of the island, is the all-day view of Rishiri, a small island to the west, consisting of a single mountain. Somewhat over five thousand feet in height, its figure resembles Fuji, though the cone is not quite so regular nor the summit so sharply truncate. Ravines full of snow extended downward from the top, across which a filmy white cloud occasionally trailed itself slowly.

A TYPICAL AINU

Primitive fishing villages lay along the shore, with many new houses of wood showing as yet no weather stain. Only lately have Japanese begun to colonize these far-away possessions of the Emperor. But they must have been excellent housewives at Onivake, for on numberless roofs lay *futon* (Japanese bedding) exposed to the fresh morning sunlight. An occasional temple showed its fine roof lines; multitudes of bright flags, each announcing the name or occupation of the dweller below, lent gayety to this little town, lying against a dark background of cedar and spruce forest. The industry, other than omnipresent fishing, is collecting edible seaweed, which is dried and sent all over Japan, even to China.

At Oshidomari, while the steamer officials transacted necessary business on shore, we lay at the foot of a high green cliff crowned by a white light-house.

Nine miles north of Rishiri is a still smaller island, Reibunshiri, and more villages. Washed by three remote northern waters, — the Sea of Japan, the Gulf of Tartary, and La Perouse Strait opening into the sea of Okhotsk, — the climate in winter is intensely cold, and the sea so rough that no steamers attempt an approach. Even in spring landing is prevented by thickly-spread fishing nets all over the bays, often far into the roadstead. One little town, Kabuka, is so exposed that even on the quiet day when the Kwanko-maru came near, so tumultuous was the surf that a large sampan heaped high with shining seaweed, collected for hours in favorable localities and now being brought to land, was overturned some distance out and all its yards of kelp treasures scattered once more into their native element. Instantly a dozen men leaped into the surf and rescued most of it.

A big boat sent out to us from the steamer agency was propelled by eight men in various stages of queer clothes and mahogany skin; one man elaborately arrayed in three separate short *kimono*, but with brown extremities exposed to chill wind and sea. After the manner of coolies, they sang at their labor, and I have written the notes of Kabuka's refrain. But voices are seldom in exact unison, and an untranscribable vocal quality makes it impossible to convey a real idea of these constantly reiterated strains: —

Japanese melodies are not easily reproduced with European instruments or voices or notation. While these are generally minor, I observed that Ainu at their work sang with entirely different musical characteristics; in major keys and with excellent rhythm, their airs were melodious to foreign ears.

In these cold regions Japanese seem hardly like themselves. Palms and bamboo belong to the national expression. But a scarlet sun was setting in a gray sea as we entered La Perouse Strait; and far in the north loomed low the shores of Saghalien, now a Russian penal island settlement, but formerly owned by Japan. Wakkanai, just below Cape Soya, was now our next anchorage, and by sunrise — Esashi.

CHAPTER XXVII

ESASHI IN KITAMI

I traveled among unknown men
In lands beyond the sea —
        WORDSWORTH, *England.*

Child of the Sun! to thee 't is given
To guard the banner of the free.
        DRAKE, *The American Flag.*

BEARS, barbarous Ainu, the Imperial Agricultural College at Sapporo, and the fine harbor of Hakodate, where men-of-war of various nationalities are apt to take refuge from the summer heats of Yokohama, — these are all that average travelers in the Mikado's Empire connect with the great northern island, Yezo.

Containing nearly thirty-seven thousand square miles, practically all forest, the number of its trees is estimated at one hundred and twenty-nine million, — evergreens in abundance, and oaks, elms, walnuts, birches, maples, and other familiar northern species, the handsome ash, and a tangle of interlacing vines.

In 1877, Mr. Benjamin Smith Lyman made the first attempt at a geological survey of the island, and many interesting facts were brought to light. There are several volcanoes and sulphur mines, also there is much coal; but speaking broadly, the Hokkaido is an unknown region, — one of the few places yet remaining where primitive nature and human nature may still be found, as rude aborigines pursue their unmolested way, and where many hundred miles of trackless forest yet await the first step of outer civilization. Nothing less, certainly, than an eclipse could have attracted to its remote wilderness at once scientific men from England and France and America, or even from the classic shades of the Imperial University at Tokyo.

In the brilliant morning sunshine the Kwankomaru pursued her tour of investigation along Yezo's northern coast in search of Esashi. The handsome young Japanese who commanded the steamer had never been there before, and the sombre evergreens, silent mountains, and gray-roofed villages on the shore afforded no distinctive landmark.

I stood on deck with Captain Kimotsuki looking at the monotonous stretch of country through a field glass, when suddenly my heart began to beat with singular rapidity, quick tears sprang, while for a moment a certain huskiness of voice prevented my telling him, and Murakami-san, calmly gazing shoreward, that I had just made out the stars and stripes, fluttering for the first time in breezes blowing straight across to Yezo from Saghalien, over the lashing waves of the Sea of Okhotsk. It does not always take war for patriotism to grow with great and unanticipated strides.

If farther confirmation of the proximity of Esashi were needed there it was off the port bow, and about two miles away,—the huge black cruiser L'Alger, which had brought Professor Deslandres from Yokohama, now awaiting the completion of his eclipse observations to return him in safety to that port.

Hardly less homelike than the American symbol was the familiar French flag with its three starless stripes; the long journey from Inland Sea to Okhotsk Sea was happily accomplished,— the welcome might fairly be described as enthusiastic.

Esashi air, evidently out of the region of heavy fog, was far clearer than in southern Hokkaido, and every prospect was cheerful, since even the prevailing earthquakes almost omit western and northern Yezo in their constant visits to the Empire otherwhere.

From July 10th until August 5th there were ten perfectly clear afternoons and four only partially shaded. The Hokkaido, in its northern portions, offers a better chance for cloudless skies than the main island, but along its southern coast fogs prevail almost constantly. Reference has been made before to the excellent pamphlet issued by the Central Meteorological Observatory, giving the observations at this season for three years past at all available eclipse locations. From a careful summing up of all results, Esashi was selected, not only by our own and the French mission, but by that sent out from the University at Tokyo. The Lick Observatory party and the English expedition chose Akkeshi, on the southeastern coast, as their location. There were thus five fully equipped expeditions in the Hokkaido, awaiting the moon's shadow to reveal truths and glories hitherto unknown.

Friendliness at headquarters had brought telegrams from the Central Government to the governor of Hokkaido, and from him to the local authorities, placing practically the entire resources of the region at our disposal, — guards and interpreters, a telegraph operator who understood English, a recently vacated schoolhouse as headquarters, a tract of land adjoining for our instruments and portable house, and every intelligent Japanese resident as willing assistant so far as possible.

The mayor's wife, a tiny lady with blackened teeth, sent vases and flowers to decorate the dining-room; the editor of a Sapporo paper (in Esashi to report the eclipse) brought gifts of petrified shells and geological curiosities — all did something. Occasionally there was a mechanical drawback — as when the Astronomer negotiated for some urgent iron work, and finally received the smith's compliments, with further information that he could probably complete one hinge and a half each day.

Professor Terao, in charge of the Tokyo mission, the official party of the Japanese government, established his camp about half a mile south of the little town and back a short distance from the beach. He was well prepared to accomplish excellent photographic work, among his instruments being an especially fine photographic doublet of eight inches aperture, by the well-known optician Brashear of Allegheny. It was constructed for this eclipse, and arrived only a few days before, having been delayed by difficulty in obtaining from Germany the finest quality of new glass needed for the lenses.

The French party was established near the western end of the town, in a large open square of land, where various tents and houses, brick piers and large instruments, made almost a little village in themselves. The outfit was very elaborate, and was intended quite exclusively for work in spectroscopy, the specialty of this celebrated astronomical physicist, who has added to the fame of the already famous Paris Observatory by his successful and brilliant work in physical research in photographing the solar prominences without an eclipse and by his discovery already mentioned of the rotation of the corona with the sun at the Senegal eclipse of April 16th, 1893. Assisting Professor Deslandres were not only the gentlemen before named as having come with him from Paris, but very valuable coöperation was given by officers of L'Alger, — Captain Boutet, commanding, Captain Le Bouleur de Courlon (who had charge of the six-inch telescope, and also of observing the four contacts), Captain Hurbin and Midshipman Dumas, in charge of the photometers and thermo-electric instruments. These gentlemen were assisted by a detachment of sailors from the Alger, and their presence quite revolutionized life in the quiet little town.* The outfit of M. Deslandres was probably the most elaborate and complete battery of spectroscopic instruments ever brought to bear on an eclipse by any single expedition. Also appliances were not omitted for pictorial photography of the corona.

Preparations of the Amherst expedition have already been alluded to, with automatic arrangements whereby electricity is made to do the work of many observers, thus extending many fold the precious two or three minutes of totality, rich with tantalizing stores of coronal wealth.

Fastened upon one great central axis, made to follow the sun by the glycerine clock, were the Lyman twelve-inch reflector from the Amherst Observatory, the Draper fifteen-inch reflector from Harvard, an Edgecomb eight-and-one-half-inch reflector, numerous object glasses by Alvan Clark & Sons, the largest of which were a ten-inch lens lent by Harvard and one of seven and a quarter by Amherst, a six-inch objective made by Schroeder of Hamburg, and a great variety of photographic doublets by the Gundlach Optical Company, and Bausch and Lomb of Rochester; and a fine lens by Goerz of Berlin. In addition were polariscopes arranged and lent by Dr. Wright of the Sloane Laboratory of Yale University, two spectroscopes from Harvard Observatory and the Massachusetts Institute of Technology, and a wheel photometer for measuring variations of intensity in the total light of the corona.

All this apparatus was individually connected with the electric commutator (invented by Professor Todd and made by Mr. Thompson for this eclipse), a slowly revolving copper cylinder full of pins each of which represented a certain movement of one particular instrument at a given fraction of a second. Each pin in the barrel had an engraved number adjacent, indicating the precise second of totality when it passed beneath

THE ELECTRIC COMMUTATOR

the circuit comb; and as ninety thousand positions of pins are possible, obviously a catalogue of each motion, the time it takes place, and the instrument to which it belongs, became a necessity.

This the Astronomer had made at sea on the way to Japan, with details of execution completely worked out, so that the whole apparatus, with an almost human intelligence, might execute its programme flawlessly.

The whole thing, most complicated to invent in all its practical working, but absolutely simple in manipulation, was set up and adjusted in time, and its working was perfect; at a touch of the electric key, plates came into place, were exposed, covered, and passed out, and new ones brought up for exposure, — all with the precision of a machine. Thus was demonstrated the practicability of applying an unlimited amount of apparatus, automatically, to the various and fertile problems of eclipse research. It is possible, with the arrangement now perfected, to take between four and five hundred pictures of the corona in two minutes and a half; and that without having to depend upon fluctuations in the nervous systems of a crowd of observers, many of whom (it is to be hoped in justice to their sense of the sublime in nature) might frequently be so affected by the spectacular part of an eclipse that routine work would suffer. Tests were constantly made, and everything progressed rapidly.

A more nearly ideal headquarters than the old schoolhouse could hardly be imagined for a practical astronomer. The long main room was made a sort of workshop for the completion and putting together of apparatus. Down one step another larger space was turned by screens and hangings into a series of small sleeping apartments for the various members of the expedition; while dining-room, kitchen, apartment for cook and assistants, a large office for the Professor in one corner of which our sleeping arrangements were shut off by a tall folding screen,— all were under the same generous roof. It was luxurious camping out. True, when Professor Deslandres or Commander Boutet called they had sometimes to be received in the dining-room, their refined faces projected against a background of shadows where hams and bacon hung dimly from rafters in the commissary department; but that was only a part of the general unusualness of the experience.

Directly opposite were telegraph and post-offices,— the former swift and reliable, the latter sure, but dependent upon packhorses to and from Wakkanai, or visits of occasional steamers.

Outside a long sliding window of the old

JAPANESE CARPENTER MAKING PLATE-HOLDERS AT ECLIPSE STATION

schoolhouse, with its protecting bars of wood, an interested circle gathered as soon as the new arrival had seated herself beside it, at the Professor's study table. Children and young girls, mothers with babies on their backs, even bent old grandmothers, collected to glimpse this strange sight. But no child was too small or of too low a class to drop a tiny curtsey when it came, with an amiable *ohayo* (good morning), and when harmless curiosity was gratified the same little figure made another quaint bow politely bidding adieu in familiar sayonara. When the gaze of every looker-on was accompanied with such well-bred manners, who could complain at being a centre of attraction?

Across the street at one of the little houses fish could be bought at certain early morning hours. Family life went on innocently in full view, and very amusing were the ante-breakfast attempts of French sailors to purchase, by a curious jargon of French and Japanese, with even an English word now and then.

But when prices or lack of mutual understanding roused their ire the resultant linguistic babel became too laughable.

One might have seen much of the village life from that sliding window alone. Women did their washing in the street, entertaining one another meanwhile by continual conversation.

Diagonally opposite was an artistic lamp-post, belonging to a neat and airy native tea-house, where pretty girls sat in the veranda, guests came and went, and rows of bright lanterns swung every evening.

The village population is composed largely of colonists from the south, attracted to these remote shores by herring and salmon fishing. For less than ten years has the little hamlet been really established; the Japanese are not fond of colonizing new regions, and only the money so easily made in spring and autumn would have lured them from their natural habitat. At those seasons the number of dwellers in Esashi rises from seventeen hundred to nearly four thousand.

The master fishermen become quite wealthy, employing from thirty to fifty men, some of whom are Ainu, in the actual labor of setting nets and bringing in the spoil. They have, too, much variety in their lives, often living at Hakodate in the winter, and taking frequent business journeys to Tokyo. Their children attend good schools, often colleges, and their houses are full of beauty and tasteful arrangement. As the potential wealth of the Hokkaido becomes more widely appreciated, probably it will not long be left to merely primeval loneliness. All these colonists, while distinctively Japanese, yet live in sufficient

harmony with the Ainu, whose primitive villages are near by in all directions.

Strolling pilgrim-beggars in dingy white solicited alms with much unmelodious music, — there were attempts at *matsuri* where in place of the gorgeous floats of Kyoto were devotees, not riding in elegance, but walking amid artificial cherry blossoms in little floorless inclosures under canopies simulating rolling cars, — a pathetic deception deceiving nobody; and more secular festivals occurred, booths were erected, plays performed, and female wrestlers contended.

My first walk abroad as the first foreign woman visitor in Esashi was a memorable occasion — to both entertainer and entertained. Chief escorted me through the principal thoroughfares, followed by an imposing procession whom intense wonder kept absolutely speechless. But at last one ecstatic small boy in dark blue *kimono* tucked up to allow freedom of limb motion recovered breath sufficiently, while marching close beside the principal performer, to produce a tin trumpet, upon which he blew vociferously, attracting the attention of all beholders. This bodyguard augmented at every corner, and the whole thing partook of the nature of a triumphal progression. Most of the followers were Japanese, but a few Ainu haunted the outskirts of the throng, with stately tread and lofty expression, apparently looking for nothing unusual, and giving no evidence of curiosity, yet never failing to see every foreign figure within range. Humbly accompanying their lords, women and children followed, — far less imposing than the men. Larger and apparently stronger than the Japanese, although not taller, the older men are actually patriarchal, with their long beards, and masses of thick hair parted in the middle, while on many faces the expression is as benign and lofty as that of a pictured apostle. Part of the walk that evening was over the pathway of clean sand spread for the Emperor's portrait. In these far and simple villages the old-time, acute reverence for everything pertaining to royalty is delightfully and solemnly preserved.

But Esashi is not really picturesque, — the wreck of a native steamer cast up on the beach by storms of the previous November, with the rough tent near by where its supposed watchers lived a more than primitive life, verged on the picturesque; but the Ishikawa-maru was not sufficiently beaten in pieces to typify that quality dear to artists.

Perhaps, after all, it was most nearly approached by a small Shinto temple close to the shore, with a neatly kept graveled courtyard and two handsome *torii*, one of fine granite. The ministering priest, an odd-looking Japanese with a sparse beard and an indifferent expression, spent an uneventful existence largely in watering handsome plants growing in vases and jars around the temple.

In the same inclosure, rising abruptly from the rocks of the shore, perhaps fifty feet high, stood a little lighthouse in which every night a student lamp burned dutifully. A narrow platform around the summit, reached by an open outside ladder, was the point from which I should draw the long, filmy streamers of the outer corona during the precious two minutes and forty seconds of totality on August ninth.

One important project was necessarily abandoned. No auxiliary stations could be established, as planned, upon some distant range of hills. The whole region was simply impassable; thick, impenetrable forests clothed every height, while scrub bamboo six or eight feet high covered all the open country. Footpaths through it from one village to another never left the shore for any distance, and no telescopes could be transported inland. There were no jinrikisha, or roads, or carriages, or *kago*, or sidesaddles, but plenty of horses; and many a mile of Kitami sands has felt the galloping feet of my rough little Yezo horse, as I traversed the country far and near, while the astronomers were adjusting apparatus, and testing plates and object glasses.

On one morning ride a small colt started with us, its mother ridden by a member of the company; but after a mile or two of racing it thought better of the trip, and returned to graze on the breezy upland moor. Coming home about twilight after a day with the Ainu, the sweetly plaintive cry of sandpipers in flocks along the beach rose familiarly as we rode at a great pace on the narrow margin of sand above high tide; and a sort of lonely quail, almost a whippoorwill note, came to us out of the woods. Through the fast-falling darkness we sped away, up the bluff, spattering through the deep mud of Esashi street, followed by the wild, welcoming cries of the little colt we had left behind.

Hokkaido horses themselves deserve a separate word. They seem to possess an abundance of good qualities which their appearance would scarcely justify our anticipating. Ordinarily they use two gaits, a short, quick trot — rather an indiscriminate sort of scramble — and a smooth gallop, rapid and comfortable. Both Ainu and Japanese are fearless and skillful riders through the narrow paths among the tall undergrowth. Largely of scrub bamboo, as already mentioned, there are acres here and there superb with wild roses, their foliage richly green like the Cherokee rose; tall spikes of burnt weed (*Epilobium*) raised familiar torches; one or two rare orchids were seen; and graceful clusters of purple nightshade were now and again turning into green and yellow and crimson berries. White chamomile, irresistibly suggesting dusty roadsides in New England, grew as large and high as marguerites, while "butter and eggs" carpeted the ground, growing flat against the sandy soil, well down to its tryst with the creeping surf.

A few deciduous trees appeared among the evergreens, their autumn coloring reported as very brilliant.

Horse-flies of scintillating green, over an inch long but not aggressive, were noticeable residents of the village, — overrun also by crows, thousands perching on every gable and ridgepole, and filling the air with flaps of dusky wings and occasional impious remarks. Hawthorne was discriminating when he asserted that crows can have no real pretension to religion, in spite of sober mien and black attire, because they are certainly thieves and probably infidels. But in Yezo they are safe from molestation and proportionally saucy.

An Ainu legend relates that in time long ago, the evil one was contending with God, frustrating his designs wherever possible. Seeing that men, his especial creation, could not live without the life-giving warmth and light of the sun, he determined to get up long before sunrise, and swallow the "lord of day" so soon as he should appear. But God sent a crow to circumvent him. When the sun was rising, the evil one opened his mouth, but a crow flew down his throat instead, thus saving the great luminary. Men therefore should ever be grateful to crows; and crows know it, indulging themselves in consequence. They feel no terror of a scarecrow, flocking near in great numbers, and even perch lovingly on its shoulder.

The morning evolutions of six crows and a black cat were worthy an eloquent description. Three on each side of her, they attacked singly and on alternate sides, her nearest neighbor pecking savagely and flying away to the end of the row when she retaliated; while the next moved up and continued the sport as soon as the cat had turned upon and vanquished the nearest crow on the opposite side. That particular pussy must have been puzzled to understand why, always worsted and driven away, the number of her enemies remained on either side unchanged. This cat for some reason had a tail, — unlike the proper Japanese species.

Early dawn, just after sunrise, and when farther naps had been effectually banished by the awkward two-steps of these favored crows upon our shingle roof, was the favorite time for official calls. A knock was followed by the entrance of our interpreter, Mr. Oshima, an able student sent from Sapporo by the governor; and following him were one and another — members of the Board of Education, or government officials, or local magnates.

With morning coffee on our part, and gifts of interesting fossils and jasper of the region on theirs, these occasions were mutually gratifying. Fortunately a Japanese *kimono* was quite full dress, which simplified matters from toilet standpoints.

We received these gentlemen in the Professor's office or headquarters, around whose walls on very convenient shelves he had arranged for safety until needed numberless eyepieces, lenses, electrical appliances, a few books, object-glasses in shining brass holders, levels, transit lamps, photographic plates, — everything one could imagine needful for an astronomical expedition.

During one of these impromptu ante-breakfast receptions at five in the morning, the mayor of the town, glancing round our apartment, gave utterance to a long and elaborate speech, — duly accompanied by low bows and friendly smiles, — evidently the daintiest of oriental compliments. In effect it was that in these shelves the children of the school had been wont to keep their shoes in former days; and that he hoped a sort of reflex action from the wonderful objects now filling the same spaces might extend to every child whose straw or wooden clogs had once occupied them, imparting to each something of the devoted scientific spirit now animating the "famous men" who had come so far to see a sublime celestial spectacle.

A leading citizen of Esashi, Mr. Hiroya, had an airy house facing the sea, which every night was gayly illuminated by hundreds of paper lanterns swinging in rows and loops along the front, and he invited us to an elaborate dinner the evening after my arrival. His pretty little bride sat slightly apart, exquisitely dressed in gray silk with an *obi* of richest brocade, smiling and looking like a picture against the background of fine *kakemono*, handsome *hibachi*, and bronze vases.

Japanese cooking shows many grades, and on this particular evening everything was deliciously cooked and entirely palatable to foreign taste. Possibly, however, I should except one delicacy in the shape of a black shell-fish, a sea cucumber perhaps, which the other guests seemed to take with avidity. Captain Kimotsuki, Professor Terao, our official interpreter Oshima-san, and a number of others were present, among them a gentleman formerly governor of a northern province, containing many Ainu villages. He en-

AINU ABOUT TO DRINK SAKE

tertained us by clever imitations of certain Ainu habits in eating, drinking, and holding intercourse with guests. He not only speaks their isolated language, but is personally acquainted with every individual of that nation in the vicinity; and as he kindly volunteered to take me to all the houses within riding distance, here was a solution of the vexed question of personal approach to these shy people, — perhaps also a solution of Professor Morse's problem, the collection of Ainu relics. When our kindly host, and his servants with lanterns, conducted us back to the schoolhouse camp, visions of eclipses and Ainu, telescopes and weaving outfits, horseback rides and treasure boxes mingled in new association invitingly in the future, and to the rhythmic beat of the surf dreams continued the picture.

CHAPTER XXVIII

IN AINU LAND

With grave faces turned toward oblivion.

SPECIAL steamers and men-of-war on the sea, and cheerful eclipse camps on shore, brought a surprising summer to the northern coast of Yezo, and the innocent Ainu will probably date future history from this peaceful invasion of foreigners. It is a happy thing that some spots are still left on this fair earth where modern enterprise and cosmopolitan life can still afford astonishment.

Among the books so constantly written upon Japan in all aspects, numbering not far from a thousand, little in proportion has been put forth in English relating to the Ainu.

Ethnologists in the Orient are largely divided as to whether the aborigines of Japan should be called Ainu or Aino, and there are strong reasons in favor of each form, both of which are used by different Japanese authorities. Upon inquiring of several prominent chiefs of the nation as to how they called themselves, and which name they preferred, the answer was "Ainu" invariably, with distinct emphasis.

The people of this race would naturally be described as "hairy," even their limbs and bodies being often quite thickly covered, yet some accounts of this characteristic have been exaggerated; and the number of hairs on a square inch of an Ainu's head is said not to exceed that upon an equal surface of a European's. The illustrations show them as not unlike the bearded peasants of Russia; certain ethnologists hold that they are probably members of some branch of the Aryan family, others that they are akin to the Eskimo.

Gentle, and subservient to the conquering Japanese, it is evident that the Ainu formerly held more egotistic views than now, even fancying themselves the centre of the universe, as shown perhaps by an old national song: "Gods of the sea, open your eyes divine. Wherever your eyes turn, there echoes the sound of the Ainu speech."

Learned discussion is still in progress among Japanese scholars as to a probable *Koro-pok-guru*, or race of dwarf pit-dwellers, "people of the hollows," who may have lived before, or partly contemporary with the early Ainu, and of whom traces are supposed to remain in various localities. Ainu themselves insist that they once fought and exterminated these people. And to the end of the twelfth century constant warring between Ainu and Japanese went on, evidences of struggle still remaining throughout the empire. Arrowheads and stone axes are found in many parts of Yezo, and in shell-heaps are bones of animals, pottery, and bones peculiar to the Ainu, who would themselves be no farther advanced in civilization than the stone age were it not for the ease of obtaining Japanese knives and swords. Their primitive utensils of bark seem to serve them as well as more elaborate implements. To an ethnologist Yezo is full of interest, from prehistoric pottery, evidence of pit - dwellings and problematic *Koro-pok-guru*, to present habits of Ainu life.

Gradually driven through ages from the south to Hokkaido, the Ainu are among the few races yet retaining in this over-civilized age an utterly unspoiled simplicity. Their actual beginning has never been satisfactorily traced, but they certainly were in Japan before the present race of Japanese had arrived, and many names clearly originating in the Ainu tongue are still retained all over the kingdom.

The oldest of Japanese books (the *Kojiki*, or "Records of Ancient Matters"), written in 712 A. D., has this characteristic sentence: "When our august ancestors descended from heaven in a boat, they found upon this island several barbarous races, the most fierce of whom were the Ainu." Whatever they may have been at that remote epoch it would be difficult to imagine a more amiable nation than the few thousand present remnants of this once numerous people. Yet they are barbarians pure and simple in spite of their gentleness, an interesting folk-lore, and the practice of considerable ceremony and forms of etiquette upon certain occasions. They have no literature, no written language, and their arts are the simplest. Contact with cultivated Japanese for hundreds of years seems to have taught them little or nothing — but extreme docility. Full of a sense of kindly hospitality, they have no ambition, and no apparent capacity for mental training. It is said that the descendant of a certain Ainu prince, or high chieftain, is now perfectly content to black the boots of an American in Sapporo. If a genuinely strong, forceful leader were to appear in the race, he might arouse them. But they have no great men. Attempts at education seem to last only during the process. Returning to their own villages, they lapse into their former state, or a placid forgetfulness.

My exploring expeditions to Poronaibo and other Ainu villages near Esashi began at once, in a method quite primitive enough to accord with surroundings. The good ex-governor was indefatigable. Giving most generously of his time and personal influence with these retiring people, as well as his skill in speaking their language, my facilities for acquiring an unusual acquaintance with their curious habits were exceptional. Casual travelers visiting more accessible Ainu villages in the south of Yezo with an ordinary Japanese guide see little of their striking race customs; but coming with their especial friend and master, I was treated more as an honored guest than as an inquisitive stranger full of doubtful intentions. Everything which might be of interest was joyfully brought forth. The fact that for the first time a foreign woman was within their borders excited much curiosity, and in all the villages they were no less glad to see me than I was to study their strange implements and habits. So with exceeding good-humor, communication made easy by my helpful friend, our mutual ethnologic studies progressed nobly. I think it was a California paper which remarked some months later in commenting upon my unique journey that probably I was quite as much of a "freak" to the Ainu as they could be to me — undoubtedly true, but a somewhat unvarnished statement.

There were drawbacks, however, to protracted calls upon the Ainu, for both personally and in their houses they are quite as dirty as the Japanese are phenomenally clean. Bathing is unknown, and their dwellings are dark, uncomfortable abodes, and far from fragrant. Each has two small holes for windows, — one east, the other south. Ainu know the points of compass, and some writers have insisted that their houses invariably face in one way. But I saw numbers facing in a variety of directions, — east, north, and west. The east end of the house and its window are sacred, and outside is a row of poles upon which the master of the house has stuck the skulls of animals killed in the hunt, among them many *inao*, or "god-sticks" as offerings to numerous deities whose aid is so constantly invoked. *Hale o Keawe*, or Hawaiian tomb for the bones of chiefs, had its outside fence of idols, twelve being set in pillars in a semicircle around the southeast end, — a curious similarity in the customs of widely separate nations.

The raised part of the floor, as in ordinary Japanese houses, has a square or rectangular hole, where during my visits fagots were always burning, — long sticks, stretching out over the floor, burning at one end, until short enough to lie wholly within the fire-hole, whose left side is reserved for the master of the house.

Smoke, although supposed to find its own way out of a hole in the roof, seemed to prefer loitering among rafters and beams, — the hanging medley of household possessions and drying fish above were draped deep with soot. A jumble of domestic *débris* usually lay in corners and around the sides of the room, and always piles of elm fibre (*atsu*) ready to be pulled apart into threads and woven into the coarse cloth (*attush*) worn by both men and women. This wood-fibre is obtained from two kinds of elm, *Ulmus montana* (*ohiyo*), and *Ulmus campestris* (*akadama*). It is pulled from the standing tree, started with blows from short knives carried by the men, and peeled off in a strip perhaps a foot wide and often twenty feet long. The *ohiyo* is laid in pools of water exposed to the sun, where the bark soon separates from the wood-fibre proper, which is then split into ten strips, and dried slowly to prevent its becoming brittle.[1] The strips are afterward still farther split into threads not over an eighth of an inch wide, the various threads tied together, and wound into balls, five or six inches in diameter, many of which were conspicuous in every Ainu house visited. The women weave the thread into durable cloth about the width of native Japanese material, in pieces over thirty feet long, or about six and one half times the length of the extended arms. Such a strip occupies in the weaving three or four days; and the garment, when ornamented with indigo blue Japanese cotton sewed on in fanciful figures, is far from unpicturesque.

Ohiyo makes a brown and reddish cloth, *aka-*

---

[1] The *akadama* is chewed instead of being soaked in water; otherwise it is treated in the same way as the *ohiyo*.

dama bright tan; another cloth made of *urtica* fibre is only used for burial purposes.

I called at one house to see a very old man. The roof was thickly thatched with scrub bamboo, and within lay a middle-aged man sound asleep upon the floor, with one arm thrown over his face, his bushy hair and beard making a weird framework. Two or three shy children were eating rice near the fire-hole, over which was suspended an iron pot, full of an indescribable stew, bubbling vigorously. A pretty young girl sat sewing ornaments of dark blue Japanese cotton upon an elm-fibre garment; and an older woman, barefooted, with hair cut very short behind, was curled in a tiny heap, looking up at me from under her arm with eyes as bright and wondering in their soft darkness as those of some shy and startled forest animal.

Family treasures, as usual, were piled around the room in chaotic masses, conspicuous among them, as everywhere, several *shundoku*, round boxes with four feet, of old Japanese lacquer, in which everything of most value is kept, and which the owner will part with last, if misfortune overtakes him. Frequently handed down through generations, an Ainu not fortunate enough to inherit one will often work a year to obtain such a highly prized case.

An ancient legend relates that nearly a thousand years ago their hero Yoshitsune, brother of the Shogun Yoritomo, in escaping to the Hokkaido from his enemies, took refuge in one of these lacquer boxes, miraculously enabled to receive him, and was conveyed away by a loyal adherent to a place of safety. This is often given in explanation of Ainu devotion to these receptacles; and also of the holes in the lacquer supports — through which cords were said to have been passed, thence across the shoulders of that "faithful one" whose back received the precious burden, — a widespread fiction. *Kakemono* representing Yoshitsune are brought out on feast days and reverently hung.

At length through the low doorway approached the old man we had come to visit, but the room was so dark that his fine face could hardly show in detail. He was an impressive figure, with a magnificent brush of white hair and beard.

But oh! the smoke and odors; soot, close air, dim light, huddling family; the mental as well as physical atmosphere was stifling, and I was forced to seek the intense relief of a full breath of outer oxygen, and sunshine. Emerging, the first object my eyes happened to fall upon was the French cruiser lying off in the open roadstead of Esashi. Such are the sharp and immediate contrasts in this interesting world, — on one hand an epitome of high civilization, on the other, Ainu huts and a

OLD AINU CHIEFTAIN

near-at-hand study of an aboriginal race now rapidly dying out from sheer inability to maintain itself in the face of a more brilliant nation.

One night a dinner-party upon the French man-of-war, — the next morning a visit to a primitive hovel within plain sight, where books had never been heard of, where furniture is unknown, where lives, sleeps, eats, weaves, is born and dies, upon the floor around a boiling pot of dreadful herbs, an entire family whose one relief from intolerable monotony is the occasional bear-killing and feast.

Salutation between Ainu men is elaborate and exceedingly respectful. Stretching out their hands, the fingers are allowed to pass softly back and forth along the palms for some time, during which verbal greetings and best wishes are exchanged. Stroking their long beards slowly is the part most obvious to a foreigner; while a gentle and inarticulate sound is made in the throat, intended to convey consideration and appreciation. The formal salutation sometimes lasts but a few minutes, though often much longer. Women indulge in very humble greeting to the men, part of which consists in rubbing the upper lip under the nose with the forefinger. Preliminary motions having been made to attract a man's attention sufficiently for him to indicate that she may proceed, she waits his invitation to speak. When a man is met out of doors, women always step aside to give him room to pass. But with all this humility, although they do all the work with constant industry, and even the consolations of the most primitive religion are denied them (for women are not even allowed to pray since they are generally supposed to possess no souls), nevertheless, an angry woman is one of the things most dreaded in Ainu land. The variety of bad names at her command to call the offending person is stupendous, and the number of adjectives with which she can heap abuse is really startling. She does not scruple to make faces and otherwise annoy and frighten whoever may have incurred her anger; and the lords of Yezo are terribly afraid of a woman in this state of mind, for there seems hardly any end to the vindictive performances with which she will afflict a man who has displeased her, especially if he be her husband. The very worst thing she can do, however, is to hide his "god-sticks," or destroy them. The deities can hardly be supposed to discriminate as to the person making away with the sacred symbols, and a man who neglects his *inao* becomes an outcast; the gods being supposed to desert him, men follow suit.

Women, continually repressed and allowed no part in religion, probably sometimes become so reckless as to fear neither gods nor man, for suicides among them are not uncommon. After early youth they are by no means to be compared with the men in fine appearance. Many girls are handsome, but the women of middle age are characterized by a stolidly dull expression of indifferent and weather-beaten resignation.

Long ago, in the first days when travelers caught sight of Ainu women, it is not strange that they were described as wearing mustaches, since, from a short distance away the heavy, blue-black tattooing around the lips gives exactly that unlovely effect. The process of producing such mouth-decoration is described as exceedingly painful, but the Ainu women have borne it heroically, sustained by their happy certainty of a beautifying result. Horizontal slashes are made with a sharp knife, crossed by slanting cuts very close together and subsequently opened wider. Coloring matter, made from the soot of birch wood scraped from the bottom of an iron kettle, is then rubbed in unflinchingly, and afterward washed with water in which ash bark has been soaked, to produce an indelible stain. For two or three days the lips are so swollen and sore that moving them, or attempting to eat, is almost impossible. But when once healed, imagine the satisfaction of emerging among one's friends and enemies, decorated for life! Many women have their hands, wrists, and arms similarly treated, showing shadowy rings and bracelets in every available spot;

and I saw a few with heavily ornamented foreheads. Young girls are attractive, for the dismal tattooing was forbidden by the Japanese government about eleven years ago, and while not absolutely suppressed, it must be done surreptitiously, and is far less frequent than formerly. Their clear brown skin generally shows a warm russet red in the cheeks, and beautiful dark eyes are shaded by long and thick eyelashes. In the younger generation, too, the luxuriant black hair is often simply coiled instead of being cut in the strangely awkward native way, perfectly short at the back of the head nearly halfway to the top, and standing out thickly on each side like an overgrown hearth-brush. A blue and white Japanese towel is sometimes rather artistically twisted around the head. It has been reported, though I did not notice this, that wives of chiefs wear a string wound six times round the waist, those of common men but three. Ainu women do not blacken their teeth, as Japanese fashion formerly decreed for married women of that nation, but they have handsome teeth, white and even. Inordinately fond of jewelry of whatever material, the richest woman is she who owns the largest number of necklaces, made of large porcelain or stone beads with huge circular ornaments suspended from them, sometimes pieces of leather studded with bits of brass or German silver. The beads

AINU WOMAN CARRYING CHILD AND BURDEN

are undeniably picturesque, many of a brilliant turquoise blue, and oddly mottled ones brought from Saghalien. These necklaces are worn at bear-feasts, when everything is in gala array for the only great occasions of the Ainu year.

Of course I wished to purchase one of these characteristic ornaments; and at last I found a woman, who, contrary to the usual custom, thought she would like some money; and rather sadly, yet with much pride, brought forth a box containing five bead necklaces. She was certainly a person of great consequence; but she fingered her possessions lovingly, looking regretfully at her cherished riches, though allowing me to examine them, while she said softly in her strange native tongue that the foreign lady might take her choice. Personally she would undoubtedly have been satisfied with very little money; but an old Japanese man in the village, of much apparent authority, sent word to her that as he had originally purchased the beads before she had come into possession of them, he would tell her their exact worth. Whereupon he proceeded to estimate the value, bead by bead, making the gentle Ainu woman open her soft brown eyes in amazement under their long lashes, and causing considerable discouragement in the breast of the would-be purchaser. We came to an ultimate understanding, however, and I bore the necklace away in triumph.

When some person of high rank in the nation comes to an Ainu house, a formal and ceremonious *sake* drinking takes place. A drop is whisked off the "mustache-lifter" to the god of the sun, Chippu Kamui in the Ainu language; next, one to the god of mountains, Kimon Kamui, then the god of the sea, Atoi Kamui, to the god of Hokkaido, Mushirori Kamui, the god of villages, Kotangoro Kamui, the god of the house, Tsuigoro Kamui, the god of fire, Abe Kamui, and to the god of all, Obishida Kamui, who is included last with a comprehensive sweep of the mustache-lifter around the whole room. Only the first cup of *sake* must be thus dispersed to the reigning powers; all subsequent drops being religiously kept for the active participants in the ceremony, who may then proceed to enjoy themselves with light hearts. These carved sticks, used to lift the heavy hair from the lip when drinking, are often elaborately ornamented.

During one of my rides, a number of rivers had to be crossed, either by fording, or by a primitive boat pulled across by a rope. One village of about twenty houses was close to a stream, and as we rode directly to the ferry, in order to get luncheon at a Japanese house a mile or two beyond, several thickly bearded men followed to watch and perhaps assist the embarkation, while a handsome girl ran down to beg that we should stop on the return; for she must see the foreign lady, fearing no other would ever visit the village. A withered old crone, bent quite double, and walking with much difficulty by aid of a long staff whose curiously carved top reached high above her head, hobbled after, giving voluble directions to the men about getting us over the river. Quite different from the expression of the older women generally, her face had a keen, cunning, almost sinister look, and bushy white hair stood out on both sides as if electrified. Huge hoops of German silver ornamented her ears, and a broad brass bracelet her tattooed arm. Her mouth, too, was heavily tattooed, and she held her elm-fibre robe tightly together with one shriveled hand. Across the river her small, sharp eyes followed us, even after we had struck into a quick gallop on the beach beyond. A weird fascination hung about this odd antiquity, and fortunately on the return a hard shower necessitated taking shelter in the house where she seemed to live.

Around were grouped daughters and granddaughters, both generations with babies strapped upon their backs, Japanese fashion, all but the youngest girls showing the disfigurement of blue-black stripes around the lips. Fagots burned as usual in the square hole, and lying about it were a number of lazy Ainu men, their strong, almost prehensile toes luxuriously spread out to the blaze. The whole household made way politely for the drenched foreigner and her companions, — producing tea and sweetmeats after hats and gloves had been taken to the fire to be dried. It was here that one of the younger girls promised to give an Ainu dance; but afterward, overcome by shyness, she slipped away.

Several women were, as usual, industriously engaged in sewing upon the aprons and *kimono* of elm-fibre different figures cut from Japanese cotton; and one was weaving the woody cloth in a primitive loom quite handsomely carved. It was a strange scene, — dark room, fitfully flickering fire, idle men with their noble faces, industrious women working by the firelight or leaning toward the faint light coming in at an open door from the clouded day without, and the visitors in the midst of them, treated as honored guests yet not disturbing the family routine. Just outside, the dripping horses waited to be remounted, ready to resume their miscellaneous scramble or free, wild gallop back to Esashi, while sharp-nosed dogs with glorious thick yellow coats peered in at the door.

I found here a small man, dark and very hairy, with a gentle expression, who was willing to sell "the best bow in the village." He had actual tears in his eyes as he told me how many bears it had shot, but that now, since hunting with poisoned arrows was forbidden, he saw no use in keeping it longer, — a small tragedy in its way.

The Ainu seem to consider the world as round, yet they are quite ignorant of astronomy, and regard the Milky Way as the "river of the gods," affording excellent sport to divinities who spend their time fishing in it. I discovered that great fear is held of comets, or "broom stars." They call one lunation a month, and twelve lunations a year. In their language star is *kidda*, the sun *chipkommoi*, and the moon *kuny chipkommoi*.

Fish, and in later years rice, with a few vegetables cultivated by women, certain lily bulbs and seaweed, form their usual food, bear's meat and venison being great luxuries.

In early spring, when the deep snows of a Yezo winter are yet hard upon the ground, the mighty Ainu hunter sets forth upon the only occupation which seems to him worthy of manly attention. The favor of the gods is always asked before starting out on one of these exciting and momentous excursions, the deities presiding over mountains, rivers, springs, and fire being entreated in turn to lend aid to the enterprise. After the bear has been killed, either in its den where it still lies partially torpid until warmer weather, or just outside, having been annoyed into emerging, or in the pit where it has been decoyed, the hunters make profound obeisance to this object of their admiration. Spring-bows or traps are sometimes stretched in the woods, when the unhappy bear shoots himself. Upon returning to the village the whole scene is related in realistic fashion to those left behind, while the deities are praised for their gracious presence which brought success to the hunt, and *sake* is taken in unlimited quantities with bear's meat at the great feast. Getting stupidly drunk upon *sake* is, indeed, the chief vice of an otherwise amiable and harmless race. It is said that of the Ainu men nine out of ten are *sake* drunkards. Fortunately the women are not considered worthy to receive enough of the precious liquid to reduce them to any such state.

Bear cubs, often taken alive, are nourished and brought up by the women in the same way as very young infants. This curious fact, stated by some writers, has been as vigorously denied by others; but Esashi held many eye-witnesses to the reality of this barbarous custom. When the baby bear gets too large for a safe playmate in the house, a great entertainment is made to which guests are invited, even from distant villages; the women are arrayed in all their pomp of jewelry and beads, some going so far, it was asserted, as to wash their hands. The men put on their head-dress of shavings, and the sacred sticks of shaven willow are stuck in the hearth as offerings to the gods. The little bear is then killed in a very cruel manner, after his pardon is asked for doing away with him. Scenes of revelry follow for two or three days, when *sake* is again drunk to excess, and rioting prevails until the meat is all eaten. Then the village resumes its wonted dullness. Bear festivals, now becoming rarer, are the opera, theatre, afternoon tea, reception, and dinner-party of the Ainu.

Shooting bears with poisoned arrows has now, like tattooing the mouth, been forbidden by government. The poison with which the hollow groove in the arrow-head was filled was made from a combination of the brains of crows, ashes of tobacco, and two kinds of insects, one of them the *krombi*, a water insect found attached to sticks and stones, the other called *yonsike*. These four ingredients mixed together and allowed to decay form a strong poison. Sometimes, however, the deadly nightshade was used instead. In Saghalien aconite roots are cleaned and scraped, then sliced and pounded to powder, which is boiled and strained, boiled again, and carefully put away, perhaps in a shell. Six dead spiders are boiled, and put in another shell; and the gall from three freshly killed foxes is also boiled by itself. These three concoctions are then mixed, and the strength of the combination tested by touching it to the tongue.

Ainu implements, garments, and utensils have often, like lacquer treasure-boxes, been handed down for uncounted years. Frequently a family has but one of each article, and that highly prized, which accounts for a prevailing disinclination to sell their possessions. To buy anything from an Ainu house requires tact and diplomacy even more than that necessary in purchasing old mahogany or china from some unwilling but hesitating elderly lady on a lonely New England country road. My knowledge of the Ainu tongue being even less than my familiarity with Japanese, I left all these little amenities to my companion, only telling him that I would buy everything they were willing to sell. His persuasiveness, and the promise of unlimited *sake* besides purchase money, brought me a miscellaneous collection of Ainu robes of elm-fibre, and one of highly ornamented salmon skin, bows and a quiver of poisoned arrows, weaving apparatus, carved "mustache lifters," tobacco boxes, knife handles and sheaths, and a rude stringed instrument. He also induced them to part with other dearly cherished heirlooms; and one or two pieces of old Japanese lacquer, made for Ainu use, have found their way to a distant land, as well as more primitive utensils of birch bark. The larger part of this collection has gone to Professor Morse, and has become part of the Peabody Museum at Salem. I have, too, a wooden eating bowl, rudely carved. As it was never washed, but merely wiped out with the finger after using, it has ac-

ARTICLES GATHERED IN AINU HOUSES

Robe and shoes of ornamental salmon skin

Old lacquer treasure-chests, rice bowls, and teapot

quired a rich and polished brown surface. I do not use it for bonbons. A "deer-call" I did not find,—a bamboo instrument with skin stretched across, by which the cry of deer is imitated.

Aprons, ankle coverings, and bands passing around the forehead by which women carry heavy burdens on their backs, all made of elm-fibre cloth, I succeeded in obtaining; and still better, two "god-sticks," the *inao* mentioned before. They are not idols, but more properly offerings to the god. Maple and willow are commonly used, one end being converted into long and fine curly shavings, either pulled apart in a fluffy mass or kept in different sorts of careful ringlets. The fluffy one is dedicated to the god of fire, the smoothly curled one given me, to the god of the mountains. They refused any money for these sticks, which are made with some sort of sacred ceremony, but signified their willingness to accept a few quarts of *sake*, and of rice. These luxuries, dedicated to the god in whose honor the sticks were made, are rededicated, after sufficient time has elapsed, to the master of the house and his friends in a more practical way. The Ainu near Esashi had quite taken me to their innocent hearts, and every day some of them came with one thing and another, learning that I really enjoyed their utensils and ornaments. When an old woman appeared at the eclipse station carrying one of their greatly valued round lacquer boxes, with permission for me to buy it, I felt that I had really won their affection.

In these northern regions Ainu often possess two residences, perhaps because of the extremely rigorous climate of Yezo, necessitating greater shelter during deep snows. The Sea of Okhotsk is sometimes blocked with ice for many miles from the coast. The winter home is called in their own language as nearly as I could write it, *riya kotan*, the latter word meaning "residing place," while *riya* is the equivalent of "to pass the year," applied to the winter. Their summer home seems to have no corresponding term, but if in Horobetsu, for instance, it would be called *Horobetsu-tsui-karu*, "to build" in that town. This was told me at Esashi by Japanese who speak the Ainu language, and by an Ainu himself.

As a people they are very superstitious, and fortune-telling prevails to a certain extent, not by the lines of the palm, but in ways quite as picturesque and perhaps not less effective. After dark the fire is extinguished, and two small bamboo sticks crossed and tied together are laid before the fortune-teller, who begins to pray aloud. Before long, so an intelligent Ainu told me seriously, the bamboo sticks stand upright unaided, and are said by some of the more devout actually to dance, thus indicating that the spirit of the god has entered into them, and is quite prepared to reveal the unknown. The fortune-teller is then moved to speak their fate for others in the assembly, who keep their heads devoutly bowed.

Medicines and care of the sick are recent innovations. Formerly, when a person became ill, he simply wrapped up his head and lay down uncomplainingly to die, — the chief attempt to circumvent fate being prayers to the gods, although certain herbs, in various strange decoctions, were used for familiar diseases. Superstitious ceremonies accompanied drawing out evil spirits, and charms were given to bring back the god of health.

But when death has actually taken place, the subject is so full of horror that the Ainu wish to forget it as soon as possible. Some necessary formalities have to be endured, however. Large household fires and feasts begin, crowds assemble, the chief treasures of the dead person are brought out, and countless god-sticks are made and placed about the body and the house. Finally, the corpse is buried, and they try at once to forget the place of burial, although sticks cut in the form of a spear, for a man, are placed at the grave; but the Ainu will not tell strangers where their dead are buried, and any ethnological collection is a remarkable one which can boast a "grave-post" or an Ainu skull. Each grave is in a separate locality, far away in the forest or among the mountains, and fear of ghosts is so great that the survivors almost never visit a grave; the posts are apt to disappear soon, and the whole matter is covered in oblivion. As an Ainu stands in deadly terror of an angry woman, so he fears nothing so much as the ghost of an old woman, thought to be full of maliciousness and power for evil. A sort of belief in an individual immortality is thus shown to be inherent, in spite of the refusal to believe practically that a woman has a soul. Some of their certainties about a future existence would be of great interest to psychical societies.

Few tribes remaining anywhere, indeed, will so well repay study, yet there are few of whom so little can be known. With no written language there can have been no reliable records, and their dread of speaking of the dead is an impediment to the accurate transmission of verbal history. Necessarily the Ainu are being pushed to the wall by the keen and brilliant Japanese, and have well been said to live "a petrified life." Yet the government makes wise laws for protection of these children of the nation, and acts toward them in an altogether civilized manner. A society exists in Sapporo for their assistance, which numbers among its members several distinguished Japanese scholars, one of them Professor Nagata, an expert in Ainu matters, and one of the best historians in Japan. One result only is inevitable from the collision of two races where one is far inferior and the other is masterfully conscious of itself.

Although a late census numbers about 17,000 Ainu, a slight gain over previous years, the impression seems to be generally prevalent that they are actually and steadily dying out. Half-breeds, Ainu and Japanese, rarely survive the second or third generation. The race evidently lacks force, and will be entirely unable to hold its own in the march of nations. Bears are decreasing in number; many characteristic customs are forbidden by law, and will soon die out completely; and gradual extinction of the race will be a pathetic feature of the further development of the Hokkaido.

But sun and moon, in their inconceivable flight through space, were almost in line, the day was close at hand, and my interest in these singularly fascinating Ainu was lost in a study of clouds and weather conditions, the working of apparatus, and of celestial rather than earthly curiosities.

Summer's climax came upon Esashi.

CHAPTER XXIX

THE ECLIPSE

> To solemnize this day the glorious sun
> Stays in his course.
> SHAKESPEARE, *King John*, III. i.

FRIDAY, the seventh of August, dawned portentously, with a strong south wind and drifting clouds. It was very warm, and bright at intervals. By evening rain set in, and all night torrents of water fell on the roof with a noise like shot. Saturday brought more south wind, occasional rain, moving cloud. Once in a while spots of blue shone through — increasing the nerve tension. The Astronomer, cheerful, energetic, showed no sign that nature's vagaries and threats were disturbing him, but, constantly busy with final details, passed from one instrument to another, clear, methodical, definite. Working of apparatus was perfect; motions were made with automatic precision, all within the time limit, all without human intervention except to press a key at the start which sent electric currents through its mysterious, ramifying nerves.

Saturday toward evening the rain suddenly ceased; a fresh feeling in the wind disclosed a change to the hopeful west, bringing a superb sunset, — shreds of rose and salmon and lavender glowed against a yellow background.

During the two days' rain none of our usually multitudinous callers had appeared; but by the light of sunset a dozen or more came together, — guests of distinction in the town as well as the village officer and leading citizens.

Another elaborate speech was made, explaining that in the storm their hearts had failed them; they could not look at this fine apparatus, remembering our patient preparation, when a chance of cloud on Sunday might ruin everything; but that now in the light of a bright sunset they came joyfully, bringing congratulations upon the weather from the fishermen, who were said to know all signs of the sky; and with hopeful portents from a book of prophecy and a local oracle, interrogated at a neighboring shrine. This cheering oracle we believed the more readily as telegrams from Sapporo and from the Central Meteorological Observatory at Tokyo announced "Clear to-morrow!" In truth all promised happily.

Stars enough came out in the evening for final tests of the instruments, and everything was in readiness.

Directions for observing the eclipse had been written by the Astronomer, translated into Japanese, printed and distributed to inhabitants all along the pathway of anticipated darkness, and some school-teachers in the village were to ascend a fairly accessible hill near by with implements for drawing the corona, and with a photographic instrument lent from our camp.

Sunday dawned through a heavy shower. Sunshine succeeded : cloud followed blue sky, northwest wind almost supplanted a damp breeze from the south full of scudding vapor. And still the hours rolled on toward two o'clock and "first contact."

The Astronomer had arranged the programme of each person with exactness long before. He still kept calmly at work, giving final directions, the multitude of details resolutely kept in mind with a philosophy as imperturbable as if skies were clear, and cloudless totality a celestial certainty. Vagaries of the western horizon, the moods of wind and prevailing drift of cirrus and cumulus had no farther power to annoy or distract. Time was too precious. It remained for the unofficial member of the party to alternate between such hope and despair that nervous prostration seemed imminent. She watched the attempt at clearing, a matter of but a few hours, and still hoped it would come in time.

At one o'clock almost half the sky was blue — two o'clock, and the moon had already bitten a small piece out of the sun's bright edge, still partly obscured by a dimly drifting mass of cloud. Half after two, and a large part of the town was ranged along the fence inclosing our apparatus, once in a while looking at the narrowing crescent, but generally at our instruments, the sober faces in curious contrast to sooty decorations from their bits of smoked glass.

And then perceptible darkness crept onward, — everything grew quiet. The moon was stealing her silent way across the sun till his crescent grew thin and wan.

The Ainu suppose an eclipse is caused by the fainting or dying of the sun-god, toward whom, as he grows black in the face, they whisk drops of water from god-sticks or mustache lifters as they would in the case of a fainting person.

But no one spoke.

Shortly before totality, to occur just after three, Esashi time, Chief and I went over to the little lighthouse and mounted to its summit, — an ideal vantage ground for a spectacle beyond anything else it has ever been my fortune to witness.

A camera was propped up beside me, with a plate ready for exposure upon sampans and junks near by, to test the photographic power of coronal light. Black disks, carefully prepared upon white paper, had been distributed to a number of persons, and several others were ready on the little platform, for drawing coronal streamers.

By this time the light was very cold and gray, like stormy winter twilight. The Alger rested motionless on a solid sea. A man in a scarlet blanket at work in a junk made a single spot of color.

Grayer and grayer grew the day, narrower and narrower the crescent of shining sunlight. The sea faded to leaden nothingness.

Armies of crows which had pretended entire indifference, gazing abroad upon the scene, or fighting and flapping on gables and flagpoles with unabated energy, at last succumbed and flew off in a body, friends and enemies together, in heavy haste to a dense pine forest on the mountain-side.

The Alger became invisible — sampans and junks faded together into colorlessness; but grass and verdure turned suddenly vivid yellow-green. A penetrating chill fell across the land, as if a door had been opened into a long-closed vault. It was a moment of appalling suspense; something was being waited for — the very air was portentous.

The circling sea-gulls disappeared with strange cries. One white butterfly fluttered by vaguely. Then an instantaneous darkness leaped upon the world. Unearthly night enveloped all.

LIGHTHOUSE ON THE BEACH AT ESASHI (from a drawing by Mr. Thompson)

With an indescribable out-flashing at the same instant the corona burst forth in mysterious radiance. But dimly seen through thin cloud, it was nevertheless beautiful beyond description, a celestial flame from some unimaginable heaven. Simultaneously the whole northwestern sky, nearly to the zenith, was flooded with lurid and startlingly brilliant orange, across which drifted clouds slightly darker, like flecks of liquid flame, or huge ejecta from some vast volcanic Hades. The west and southwest gleamed in shining lemon yellow.

Least like a sunset, it was too sombre and terrible. The pale, broken circle of coronal light still glowed on with thrilling peacefulness, while nature held her breath for another stage in this majestic spectacle.

Well might it have been a prelude to the shriveling and disappearance of the whole world, — weird to. horror, and beautiful to heartbreak, heaven and hell in the same sky.

Absolute silence reigned. No human being spoke. No bird twittered. Even sighing of the surf breathed into utter repose, and not a ripple stirred the leaden sea.

One human being seemed so small, so helpless, so slight a part of all this strangeness and mystery! It was as if the hand of Deity had been visibly laid upon space and worlds, to allow one momentary glimpse of the awfulness of creation.

Hours might have passed — time was annihilated; and yet when the tiniest globule of sunlight, a drop, a needle-shaft, a pinhole, reappeared, even before it had become the slenderest possible crescent, the fair corona and all color in sky and cloud withdrew, and a natural aspect of stormy twilight returned. Then the two minutes and a half in memory seemed but a few seconds, — a breath, the briefest tale ever told.

As the beautiful corona lay there in the clouds, a soft unearthly radiance, the poetic effect as strong as if in a clear sky, the scientific value lost in vapors, it was still noticeably flattened at the poles and extended equatorially, and must have been of unusual brilliance to show so distinctly through cloud. The shape gives suggestion to astronomers as to new lines of future research.

Just after totality a telegram came from the Astronomer Royal of England, far away on the southeastern coast at Akkeshi: "Thick cloud. Nothing done."

Nature knows how to be cruel, or possibly it is mere indifference. But until, in his search after the unknown, man learns to circumvent cloud, I must still feel that she holds every advantage. On that fateful Sunday afternoon the sun,

EXPEDITION MEMBERS, AND OLD SCHOOLHOUSE, AFTER THE ECLIPSE

emerging from partial eclipse, set cheerfully in a clear sky; the next morning dawned cloudless and sparkling.

A few pictures of the blurred corona were taken, if of little practical use, and an interesting experiment for Roentgen rays seemed to indicate their presence in coronal light, — a curious result, since they have not been found in full sunlight.

But a useful and tangible outcome of the expedition is afforded by this practical demonstration that a great number of instruments can be employed in recording the corona automatically, not only dispensing with the multitude of assistants necessary for manipulating each at critical moments, but virtually lengthening the precious minutes of totality many fold.

The corona, thus safely caught, can now be laid on our tables in manifold representations, and interrogated through the months following an eclipse until the most telling questions for its next coming are plainly evident.

And Esashi had vindicated its choice. Of all the places where meteorological observations had been made, it proved the best — clouds, that is, were thinnest.

Nothing appeared upon the plate exposed to the sampans; coronal light was not strong enough to impress them upon the sensitive surface.

But the apparatus remains — from the approach of the idea in Shirakawa, in 1887, when it was roughly but accurately carried out for the eclipse of that year; the far better evolution in West Africa in 1889 by pneumatic contrivances; and the smoothly running devices intrusted to electricity in Yezo in 1896, — perfected result of three cloudy eclipses.

CHAPTER XXX

A NATIVE CELEBRATION

Whilest that the childe is young let him be instructed in vertue and lyttera-
ture.
LYLY, *Anatomy of Wit.*

Thou hast most traitorously corrupted the youth of the realm in erecting a grammar school.
SHAKESPEARE, *2 Henry VI.*, iv. 7.

Schools — they are the Seminaries of State.
B. JONSON, *Discoveries.*

IT might naturally be supposed that an American in northern Yezo would have time and to spare. The facts were that I was breathlessly hurried every day. Always there were more things to do than hours for their accomplishment. The Ainu must be constantly visited and studied, collections of ethnological articles increased, copious notes taken of all queer surroundings, calls received, horseback rides up and down the coast (when occasionally a horse would lie down while fording a stream, — merely a temporary inconvenience), and, most distracting of all, the new school-building was dedicated with elaborate ceremonies, followed by a great dinner, the whole occupying an entire day.

The Japanese are soberly and deeply interested in education. The presence in this remote region of so many members of the Board from southern Yezo for the coming ceremonies sufficiently attested that. And the eleventh of August was the great day.

There was a certain peculiarity in thinking of Esashi as an educational centre, but the doughty ex-governor, through whom I had reached the Ainu, who seemed to own the whole region, and to whom every inhabitant for miles around bowed to the ground when he passed, comes to Esashi for its advantages to his nine children. He was a Samurai retainer of Matsumæ in old feudal days. Generally he has a number of Ainu as servants, and reported that they do excellently until spring, when, being consumed by a desire to drink *sake* to excess, they become practically useless. He is a sort of feudal dignitary himself now, and had just built an immense new house, where he one day gave an elaborate Japanese luncheon. Eggs were served in six different ways; but the most remarkable dish was the roe of sea-cucumber, kept for three years and eaten with a few drops of vinegar. Pink and elastic, it was considered a great delicacy.

On dedication morning he accompanied me for an ante-breakfast walk to some Ainu houses near by, thus far omitted in my longer trips. All the way relays of Japanese children were met, in their best clothes and with clean and happy faces, starting joyfully for school, the little girls in bright *kimono* and *obi*, their hair, black and shining, ornamented with gay hairpins, the boys in white-crowned black caps, gray or dark blue *kimono*, and divided skirts. As we passed along, the friendly governor told many interesting tales of Saghalien, where he lived for several years. Curious things find their way to Yezo from that far island, and amid the constant groups of smiling children his talk grew reminiscent.

Six races inhabit those chilly shores,—Ainu, Manchurian, Kurin, Oroku, Nekubun, and Sauran. The governor had already presented me with a handsome fur rug made by Kurin women there from the fine head-fur of an animal somewhat resembling a deer, but with larger feet and heavier ankles, and horns showing thirty or forty small branches. Natives call it *tonakai*. Mothers in Saghalien suspend bark from the rafters of their huts, in which the baby is swung, a string attached from it to the foot. Although in 1857 there were over two thousand Ainu there, they have now dwindled to less than half.

Returning to camp for breakfast, a committee of officials, including the vice-governor of Hokkaido, and governor of Kitami and two other northern provinces, was found waiting to conduct the Astronomer to the school building. Very soon an imposing procession set forth for that seat of learning, surrounded by its turf fence. A huge triumphal arch of evergreen surmounted the entrance gate, with festoons of scarlet and white flags and lanterns. Dignitaries and guests were first ushered into a room with low tables where tea was served, adjourning afterward to the large schoolroom beyond, filled with boys and girls, around them numerous officials of education and government, and a few Buddhist priests with finely intellectual faces. Men filling national positions had come to remote Esashi for this occasion, an evidence of earnest ambition along the best lines.

The three astronomers, Professor Deslandres, Professor Terao, and Professor Todd, sat near the closed shrine of exquisitely smooth wood, brass ornaments, and royal purple drapery, containing the Emperor's portrait. On a corner of the platform was a minute musical instrument like a tiny parlor organ, also covered with purple. At three single notes every child rose, and, dragging unconscionably, all sang in unison the National Anthem (page 154). Japanese music avoids half-tones — founded upon the harmonic minor scale, the intervals most frequently sung are strangely unnatural, the tonic playing no apparent part whatever in the basis of any melody, ending, as many do, in mid-air upon the seventh. The singing tone, moreover, is exceedingly nasal, quite different from the gentle, ordinary speaking voice, and the children's throats actually distended with pushing and squeezing the notes.

This finished, the school principal rose, faced the sacred cabinet, and bowed. Opening the doors with dignified deliberation, he exposed the portraits of Emperor and Empress, whereat every child bent to the floor, remaining in that position for two or three minutes in utter silence. At three organ notes they slowly stood upright once more. Facing the portraits, the principal then gave allegiance and congratulation in impressive tones, while all bent low once more, and the shrine was closed.

A long box, like the case of a *kakemono*, was next produced, and, opening it, a scroll was held up, containing the Emperor's message, read aloud while all school heads were devoutly bowed.

More singing, — a piece ending on the fourth of the scale as the National Anthem does on the second, — and the good mayor, Shirasaka-san, rose for a speech. Still more singing at its conclusion, after which the vice-governor Suzuki-san, read from an imposing document, and Some-san, head officer of the Colonial Department, addressed the assembly. To all these gentlemen the school rose and bowed in turn. Happily all the speeches were short. Some others had also given a few words, but their positions did not entitle them to bows. Finally one of the schoolboys read some sort of a response on the part of the scholars,— exceedingly well too ; and another song followed.

Then the American Astronomer was called upon ; the children rose and bowed, and remained standing until he finished. Among other things he presented the school with a fine framed picture of the corona of 1878, one of the famous Trouvelot drawings, urged them to have English studied, and presented some books we already had in Esashi, promising others sent later from Tokyo. This speech was gracefully translated, sentence upon sentence, by clever Mr. Oshima. Afterward Professor Deslandres made a short speech in French, interpreted by Professor Terao, who, years before, had studied at the University of Paris.

The mayor thanked the Astronomer for his gifts to the school ; an appetizing Japanese luncheon was served, the baked salmon especially delicious, and so the new building was fairly inaugurated.

Later, toward sunset, a dinner in honor of the Americans was given at a spacious tea-house newly built near by, the stars and stripes and red sun flags draped at the entrance, the feast occurring in a large upper room wide open on opposite sides to sea and town. With much deliberation the worthies assembled, occupying nearly three hours in getting there. The cooking was exceedingly fine, the serving perfect. Maids in waiting were charmingly dressed in silk crêpe, blue or pale green, with magnificently brocaded *obi*, and elaborately smooth hair, like an exquisite picture. A wonder would irrelevantly intrude itself as to what sort of waitresses would be encountered in a "tavern" in the wilds of northern Maine, or in a fishing village of Nova Scotia, localities far easier of access than this Okhotsk shore.

Just before the feast the mayor had brought in a long strip of white satin for a *kakemono*. The Astronomer, Chief, and I were requested to paint upon it either pictures or poems. Brushes, colors, Japanese ink, and water accompanied it. Chief, of course, composed an original verse. The Professor was content with an appropriate line or two from Shakespeare, while a few rushes in one corner with her name attested the modesty of a third contributor. Professor Terao placed his personal red seal upon the strip; and I doubt not a memorable *kakemono* now adorns the mayor's residence.

But days in the far-away little town were drawing to an end. Every time I came back from any excursion a few more instruments had been taken down, a few more boxes packed, a few more gifts from the kindly inhabitants brought in, as well as prospective *kakemono* in the shape of additional strips of silk and satin and fine paper for verses and autographs. Occasionally an aged Ainu or Japanese was found awaiting my return with farther aboriginal articles which I bought with alacrity — and *yen*. In the evenings the watchman going his rounds beating two sticks to announce his faithfulness lulled us to slumber, and the final day came on apace.

Captain Boutet of the Alger had courteously invited the Astronomer and his companion, also Chief, to return to Yokohama on that famous cruiser. A special Yusen Kaisha steamer, already dispatched from the south for the expedition and apparatus, was expected daily.

On the sixteenth of August, a lovely summer day with a hint of coming coolness in the air, the Commandant sent his gig ashore for us, and truly reluctant good-bys were said, not only to expedition members, still waiting, but to the few Ainu shyly looking on from the outskirts, and to a crowd of warm-hearted Japanese who had done everything in their power for our assistance, honor, and pleasure. Accompanying the gig to the Alger was the big sampan built for the Emperor's portrait, now filled with the familiar and friendly faces of Shirasaka-san (the mayor); the lovable, big ex-governor; Hiroya-san, and others who never ceased waving so long as we stood in sight upon the Commandant's overhanging after-balcony.

But the lighthouse where I witnessed the eclipse grew smaller, and faded; the little gray town disappeared. Esashi was but a memory, sad yet dear.

Most unlikely is it that we shall see Yezo, much less Kitami Province again; but a warm spot in my heart still glows at thought of this hospitable village, encompassed by impenetrable forest, surrounded by aboriginal Ainu, and facing the far north over the uneasy wastes of the Sea of Okhotsk.

> "The crimson sunset faded into gray;
> Upon the murmurous sea the twilight fell;
> The last warm breath of the delicious day
> Passed with a mute farewell.
>
> "Above my head, in the soft purple sky,
> A wild note sounded like a shrill-voiced bell;
> Three gulls met, wheeled, and parted with a cry
> That seemed to say, farewell."

CHAPTER XXXI

VOYAGE ON A FRENCH CRUISER

Et puis, peu à peu, on vit s'éclairer très loin une autre chimère : une sorte de découpure rosée très haute, qui était un promontoire de la sombre Islande.
PIERRE LOTI, *Pêcheur d'Islande.*

Le soleil . . . n'avait plus de halo, et son disque rond ayant repris des contours très accusés, il semblait plutôt quelque pauvre planète jaune, mourante, qui se serait arrêtée là indécise, au milieu d'un chaos.
LOTI.

BECAUSE of a delightful habit of the Commandant, the Astronomer and I were enabled to circumnavigate the island of Yezo. Avoiding the same course in going and returning, Captain Boutet always varies his routes when possible, and he, like ourselves, had reached Esashi by the west coast. When twilight settled over the gray sea, L'Alger was well along toward the eastern end of the island, her black bow pointed almost due east, the little after-balcony over the water holding a congenial company,— the two astronomers and the Commandant watching the fading shores, while I sat just inside the door, in the dainty salon which with the two or three other apartments forming his own private suite Captain Boutet had devoted to his newest guest.

He has been an indefatigable and discriminating collector of fine Japanese and Chinese plates, platters, and odd pieces of porcelain, which decorate superbly the walls and ceilings of these charming rooms when in port, — all carefully packed away at sea. Still, articles enough of a less frangible nature adorn them constantly, to conceal, or at least to grace, the solid steel walls of this great war vessel.

"Automatic photography of celestial objects is the astronomy of the future," Professor Deslandres was saying, as the waves beneath the balcony grew rougher, and the three came in to the brightly lighted parlor, gay with panels and *kakemono*, "and Professor Todd is its precursor and prophet."

His interest in the Amherst apparatus had been strong, as ours in his own fine spectroscopes, and many delightful calls between the two stations had passed at Esashi. But this evening at sea was really the first quiet, unhurried, and really favorable time for talking over technical matters; and I soon left them for the little bedroom with its square window opening to wide sea and sky, the fascinating blue dragon china fittings, each a separate work of art, and the luxurious bed, compared with which Japanese mats in the dear old Esashi schoolhouse felt very hard even in remembrance.

Fog occasionally drifted up, but blue sky and

sunshine soon followed on this happy voyage, and a few hours of heavy swells necessitated taking in the balcony floor. The course was laid definitely, the hour of arrival in Yokohama announced at the start by the Commandant, whose precision of movement is proverbial in the French navy. Steadily the course was made, our exact position brought to him several times during the day. Shikotan, the big island east of Yezo, was passed, and the southwest course for Yokohama begun.

L'Alger is three hundred and forty-five feet long, and of more than four thousand tons burthen. Wholly built of steel, she carries formidable guns, and over four hundred men, of whom about thirty-six are officers, the commandant, or *Capitain de Vaisseau*, having next below him in authority another officer, whose title is *Capitain de frigate;* next, five lieutenants, and others down to petty officers.

Every morning reports of all kinds are handed in to the Commandant; for instance, that three tons of distilled water were made yesterday — the capacity for this manufacture being eighteen tons, seventeen tons being now on hand; that yesterday so many tons of coal were used, leaving a definite number still in her bunkers. Since the Alger can carry many more than now remain, coaling must be done at Yokohama, before the voyage to Nagasaki. Toward that favorite port Captain Boutet says his engines beat joyful time, repeating in their throbbing, "Nagasaki, Nagasaki, Nagasaki" in quick iteration, while if the orders are to proceed to Korea, sadly, in funereal time, the machinery reluctantly grinds out "Chemul-po — mul — po" to a dirge-like rhythm.

Reports upon provisions were made daily, — the amount of wine remaining; of tafia, a sort of brandy from sugar-cane; of farina, which includes many cereals; and "divers." Certain figures, one day standing eighty-four, meant that so many meals (two each day) with wine remained; forty-two (one each day) with brandy, twenty-two of cereals, thirty-five of biscuit and thirty-four of miscellaneous articles. So it was quite plain even to the uninitiated that supplies must be laid in at Yokohama, if amounts for three months, the Commandant's rule, be carried.

Illness of any one on board is at once announced, — an officer having injured his knee was reported, while I sat there, two "petits" officers and six men being already ill, — nine in all. More interesting was the report of culprits which the Captain amiably allowed me to read. Three men were undergoing punishment, the first "Numéro, 51 B, Nom, S——; nature de la punition, B. justice 138; nombre de jours, 5; fin de la punition, 22 Août; motifs, Réclamation mal fondée et occasionner du désordre dans la batterie." Another is punished during five days, because of striking "brutalement un de ses camarades sans motif;" and a third for "negligence dans son travail et reponse inconvenante," during four days.

Each watch has a lieutenant in charge, accompanied by a midshipman (*aspirant*).

Elaborate tables of exercises are made for every hour of the morning and evening, and each day of the week, for instance: "Exercise général de manœuvre," or "Exercise général ou ordinaire du canon en alternant successivement" — these being from 9.30 to 10.30 on the mornings (*jeudi et vendredi*).

It was curiously interesting to look over these tables, and read that *lundi* the sailors get out their clean duck, look it over and mend it; *mardi* brings inspection by the captain of "materiel," in other words of guns, muskets, metal columns, brass, and for assurance, no less, that each man in charge of its condition is at his appointed post; if everything is satisfactory he has an extra ration of wine, — if not, his allowance is reduced one ration. *Mercredi*, boiler inspection, and that of knives and forks and other utensils of sailors' tables. Eight men at each table have every week one of their number appointed to see that

A "HAIRY AINU"

all things shine duly; he too is rewarded or punished according to their condition.

Jeudi, one sort of inspection goes on; *vendredi* another, and *samedi* sees general cleaning and brass polishing for a shining *dimanche*.

The first Sunday in the month the Commandant tests the men with regard to arms and place in battle; the second, one hundred and ten men with muskets are landed; the third, inspection as to their condition of four different companies, one hundred men in each; and on the fourth the same, with sailors manning boats to show their skill in rowing and general alacrity.

Our own war with Spain has made the public more or less familiar with routine on men-of-war, through numerous newspaper articles; and we know, too, the latent force and splendid energy of officers, ready to spring forth at a moment's notice in mastery of every situation, perilous or desperate; but life to the commander of a war vessel is certainly no sinecure, even in times of peace, as shown even in the small bits of routine kindly told and shown me by our host, the delightfully courteous Commandant.

Indeed, if perfect system makes his own part seem full of grace and ease and luxury, he holds no less every movement of the huge cruiser and its occupants in his hand for every moment of every day. Yet his life seemed, in those peaceful waters, as ideal as that of his guests, — beautiful quarters, perfect service, an elaborate menu, an autocrat unquestioned. And better than all, the gentle heart, exquisite courtesy, and æsthetic taste which make all life worth while.

CHAPTER XXXII

HOMEWARD BOUND

> O'er the deep! — o'er the deep!
> Where the whale and the shark and the swordfish sleep, —
> Outflying the blast and the driving rain, —
> BARRY CORNWALL.

GRADUALLY the Coronet party again assembled on their beautiful home. We were the first returned wanderers — soon followed by the Captain, Mrs. Captain, and others of the "unscientific contingent," disappointed to have found the time too short for reaching Esashi before the ninth of August, but partly consoled by the beauties of Miyanoshita and Nikko; last of all, the expedition members, unexpectedly detained at Esashi several days, as the special steamer had been caught in fog on its way northward.

Much hospitality on board was resumed immediately, — tiffins and dinners to the Astronomer Royal of England, Professor Turner, and Captain Hills, Professor Deslandres, Captain Boutet, and others; while dinners on shore to and by the various astronomers, interspersed by dancing and dining on the men-of-war, followed in quick succession.

Professor Turner as extempore poet shone in a new light. A guest book having been presented for his signature, he retired to a quiet spot on the Coronet's deck and soon produced the following impromptu lines : —

>Astronomers we,
>One, two, and three,
>(*Ichi, ni, san,*)
>Came to Japan,
>Came for eclipse,
>Sailed in six ships,
>Trained in six trains,
>Suffered from rains,
>Ice, fog, and dew,
>Hot weather too,
>Oft dry with thirst,
>But what was worst,
>Cloud interfered,
>No corona appeared.
>Some compensations,
>Coronet's rations!
>Coronet's smokes!
>Coronet's folks!
>So the best of good wishes,
>And now home, o'er the fishes.

The "edibles, bibables, and fumibles" of that fair craft, deservedly celebrated, are not always so immortalized.

Another interesting entertainment was given us by a Japanese friend at the Maple Club; and the famous drive to Mississippi Bay was taken, where the rice fields, now in a state of lovely ripeness, showed full and graceful heads, bending with a nation's nourishment. Some one announced in passing, that very poor Japanese parents sometimes give their children partly cooked rice, that by its subsequent swelling their growing appetites may, for a time, be kept at bay!

Odds and ends of pleasant sight-seeing or business were finished; and suddenly out of the intense heat one cool evening descended, suggestive of approaching autumn, and farewell.

Mr. Christie departed for England eastward on the Empress of China; Professor Deslandres went to Nikko, waiting for cooler days to begin his homeward trip by way of India and Suez; L'Alger swung loose from her buoy promptly to the moment of Captain Boutet's intention, steaming impressively away through the breakwater and bound for Nagasaki, while farewells waved from her bridge and quarterdeck so long as figures could be distinguished.

Native papers published excellent accounts of the eclipse, one of them, given below, having been translated by a guide, — not the "famous" Okita. So far as I have been able to judge, Japanese characters give very definite meaning to those who read them, but unless translating is done by a scholar, it becomes vague in the change. This guide used verbal English very well: —

"The Professor Terao sent by the Imperial astoronomical house and among foreigners American Professor Tod and party, French Parisian latitudinal bureau's Mr. Drandol and party have established the looking and surveying places here. . . .

"The all expenses to perform this object is to be delayed by rich Mr. James as the plan was made by private of individual.

"Also Mrs. Tod being an astoronomer, and coming together with Mr. Tod and helped the work to take four more Americans, herself as engineers.

"Besides the above party the photographer Ogawa also followed taking two his men.

"The machinaries has been invented by the same Professor and its principal object is for taking to the photograph the present sight of the Eclipse by moving the machinary by the action of electricity. During the time of eclipse per every two minutes 150 pieces as many, and 24 or 36 as little would be expected to be taken, so that altogether 4-500 would have been supposed to be taken in the last.

"The machinary being composed to change the direction by the same advancing rate as well as the earth revolves and there is no necessity to move the machinary's position during the eclipse, so much so conveniently arranged having had good result on several trial. . . .

... "In the evening of the 8th the cloud got clear up, gradually, and all astoronomer felt much easier, but on the 9th from the dawn, the small rain began to fall but sometimes the sun seemed through the thin part of it, while we passed the before noon with a glad and sorrow.

"About half past one clock the sun began to get waned from right side and about half past two it reached to the last part from doing so and the heaven and land became dark and showed quite night sight. As we heard before, the flying birds got much astonishment and made a great confusion to return to their own nests, and showed a special sight. . . .

"The plan and pain with each surveyors during the past a month being brought such sadful result and nobody can tell how much those astoronomers caused the distress for hopeless end like that.

"Mrs. Tod came from far place to help her husband's work, and during the time of so many days she has tried to do her best through day and night, but the weather prevented her will, and she has forgotten herself to cry out, and we ought to think about such learned lady's heart."

During the last days, frantic desire to purchase final Japanese presents, and by no means to forget this, that, or the other article, or person at home, surged onward like a tidal wave. But

Yokohama does not shine in comparison with Kyoto for shopping, — with marked exception in the case of the beautiful vases of Makudsu Kozan, sometimes called "the wizard of Ota," whose famous kilns are near the city; and perhaps one or two other celebrated places and artists. Chinese tailors and shirt-makers were driven quite wild by the sudden influx of orders. Every man discovered the necessity for several full suits, and affable Ah You spent most of his available time on the Coronet's deck, untying innumerable purple silk handkerchiefs containing coats for trying on; or in pinning them upon the happy if perspiring prospective owners below, — or in being paddled back and forth in a sampan. I cannot conceive that he did anything else whatever during those last days. He was as much a part of the yacht scenery as quartermasters and awning. And a certain shirt-maker, Yamatoya, hardly less. Really, because people go to Japan to observe an eclipse is no valid reason why they should not clothe themselves extensively with fine Oriental bargains. But

"Shining and singing and sparkling glides on the glad day,
And eastward the swift-rolling planet wheels into the gray."

A final reception on board the Coronet, never so fairylike with flags, lanterns, and groves of bamboo, and the day of homeward sailing dawned.

We aimed to clear the moorings at colors, but, detained by a number of calls, it was nearly nine o'clock when we started, Fuji dimly brooding, and slowly swung off as our sails filled, homeward pennant streaming, down the lovely bay. Passing the flagship, which has since made such quick history for herself, all her white-clothed sailors drawn up forward, our friends on bridge and quarter-deck were waving caps and kerchiefs; salutes rang out, colors dipped, the band played "Home, Sweet Home" and "Auld Lang Syne." And while distance widened, as freshening breezes caught added sails, the familiar strains of "Nancy Lee" floated off to us, etherealized by distance, this delicate compliment from the Olympia being the last sound to reach the Coronet from Japan's domain.

This voyage would see no stops at tropical islands, no volcanoes, no lawn teas,—but the shortest possible great-circle course through the northern Pacific to the Golden Gate, a lonely waste which with most favoring breezes could not be traversed in less than three weeks, and was likely to take much longer. A month absolutely without news of the world, telegrams and letters powerless to cheer or annoy,—a month alone with immensity. What better chance to make acquaintance with that stranger too seldom met,—one's innermost self?

At first outer conditions were quite different from anticipation, — light winds, then a gale, followed by calm, with smooth blue sea; suddenly another fierce blow, all sail lowered, the yacht "hove to" with such tremendous seas rolling that no amount of guards were effective to keep dishes on the table or guests in their beds; even the Captain landed suddenly on the floor one night during the brief interval of sleep he allowed himself. For on this homeward voyage he took regular watch, in turn with sailing master and mates.

At the 180th meridian, picking up a second Wednesday and reaching the same hemisphere with our friends, the seas too seemed changed, running high but in our own direction, slipping heavily beneath from the stern, while fine winds urged us forward. Many an inspiring day followed, — shaded gray skies with an occasional sun-gleam, now and then a streak of rich blue showing through layers of soft cloud, — the sea gray and green, black in its shadows, breaking white on every crest, and hurrying eastward impetuously, faster than we could race. Yet like a bird the Coronet flew over the uneasy wastes of endless water, lifting her delicate nose scornfully above the rollers, and taking few seas aboard. Fortunately there was little fog; but there were gorgeous sunsets, and one sunrise was a rose-pink pile of cirrus, deepening to ruby. Flying meteors at night; showers chasing each other blackly around the horizon; a great, impressively moving waterspout; porpoises leaping, and our old friends the goonies following as usual, flying six thousand miles and knowing not fatigue; sometimes a white shag, and a beautiful white bird like a pigeon, its little scarlet feet tucked up beneath, seeming inclined to alight, but thinking better of it; whales blowing, even lifting their huge bulk high above the water — with these diversions the days rushed on.

Another of Big Jim's stories was recalled by the whales, — he was no longer on board, having been left behind at Yokohama from circumstances over which he had not full control, so that he existed for us but as a memory. His tale was to the effect that he once harpooned a whale, which immediately set out on a mad journey, dragging the boat after him. "Why," said Jim, "he pulled us so fast though the water that as you looked astern we had left a clear tunnel through the waves a mile or more back."

A weary little land bird like a song sparrow fluttered to the deck one day, — presumably from the nearest shore, there the Aleutian Islands, more than three hundred miles northward. But fatigue, hunger, chill, thirst, or fright proved too much for his delicate life. Resting on the waves, another land bird was passed, which only looked at us with bright inquisitive eyes as we sped past at ten knots. A squid once came aboard under protest. Indignant at his sudden stranding, he proceeded to cover the deck with particularly black ink.

One morning the whole sea was alive with exquisite spots of radiant blue fire, both on the surface and far down into the water. For two days this remarkable sight continued, though no one was able to identify the startling little creatures so royally arrayed. They were evidently crustaceans, their color thrilling, iridescent, phosphorescent, flame-like.

Bottles, tightly corked, containing each a record of date and exact latitude and longitude, were thrown overboard on alternate days, that by their drift and possible subsequent landing additional data might be secured for the Hydrographic Office regarding the direction and velocity of ocean currents.

Sealskin coats and sea-rugs were much in evidence on this northern voyage, with brisk walks, and afternoon tea by the open fire; while, dinner over, Beethoven and Bach and Chopin filled the evening hour.

Shanties, too, continued, several new ones taking their places in the yacht's forecastle repertoire, among them —

"ROLL THE COTTON DOWN"

From Yokohama we're homeward bound,
  Roll the cotton down,
From Yokohama we're homeward bound,
  Roll the cotton down.
2 And soon we'll be in 'Frisco town,
3 And as we leave Yokohama behind
4 We'll try to make the fastest time
5 And beat the record as home we go;
6 It takes a Yankee yacht to do so.
. . . .
7 So pull, my boys, from down below,
8 For up aloft the sail must go.
9 I thought I heard the chief mate say
10 Another pull and then belay.

Toward the end of September, when superb weather came on, with sparkling blue, foam-capped sea, high cirrus cloud and northwest winds, the Coronet fairly leaped over the waves.

Showers still haunted the horizon, and one evening as the moon emerged from cloud, a perfect lunar rainbow brightened gradually until even the secondary bow came forth in shadowy color, — an exquisite sight, elusive, fairy-like.

East winds, cloud, and high seas took their turn before the coast was sighted, with reefed sails and tons of water sweeping the deck. In the night a blow might strike the vessel's bow until she trembled, — then the swish and rush of chasing water along the scuppers, like a huge but temporary mill-race. In spite of reefed sails we surged onward, gleaming foam thrown off from every side, the great, gray, mysterious sea heaving and trembling in dim obscurity in all directions.

During one of the last days came the sole cry of "Sail ho!" on the entire voyage, and a bark was seen hastening off in rain and mist on unknown errands.

Late in the evening of October 1st a faint whistle sounded through the fog; and soon after midnight, the weather clearing unexpectedly, we were called on deck for a moving spectacle. All the stars were out and a waning crescent moon; and just ahead, the intense brilliance of the Farallones light, our bow pointed directly for its radiance.

No longer could the faithful owner and Captain of the Coronet be gayly termed a summer yachtsman; he had fairly earned his title of skilled and experienced deep-sea navigator, if only from this splendid course through trackless waters of the northern Pacific. The two courses from San Francisco to Yokohama and back, as shown on the chart reproduction, make a pretty smooth navigation curve, counting ten thousand eight hundred and seventy miles, — the homeward voyage being within fifty miles of the shortest possible course.

el.

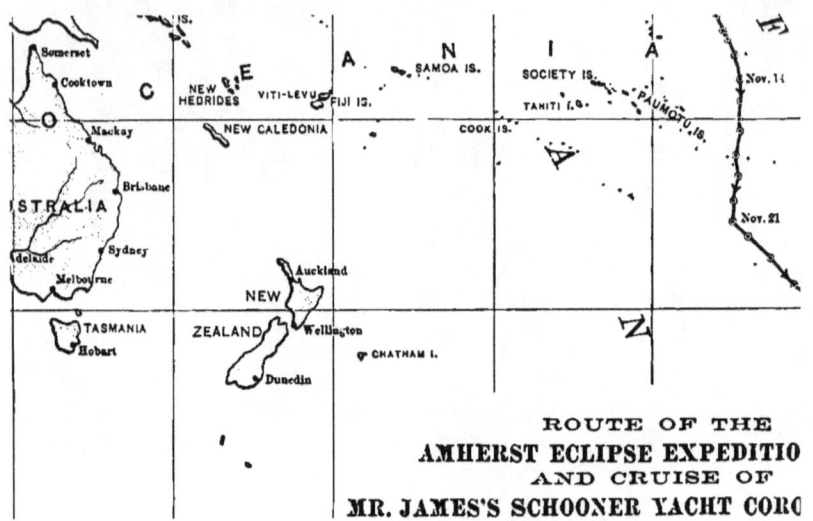

Heartfelt congratulations from every one on board to the trusty navigator, who bore his honors so modestly, — while coffee and sandwiches by the blazing fire at 1 A. M. celebrated this triumphant land-fall.

Before noon the cabin was buried deep in long-stemmed crimson and yellow roses, fragrant violets, carnations, maidenhair ferns — the vases were too few to hold them; we lunched and dined in a bower. Friends poured in, reporters poured in, invitations poured in. The curiously brilliant colors upon hats, the peculiar sleeves, all so different from things prevailing six months before, were oddly interesting. Hills, vividly green in April, now showed sober brown. Sunshine was continual. The great cliffs of the Golden Gate were more superb than we remembered them, the miles of pathless white sand-dunes as mysteriously, weirdly attractive as before.

But alas, and alas! good-bys to the Coronet were creeping nearer and nearer. What though the luxurious private car Buenaventura awaited us — farewells to Captain Crosby, faithful Andrew, and the sailors, three dips and a salute as we left her side extinguished all emotions other than unmitigated homesickness, and genuine regret (on the part of one or two) that we might not go with her round the Horn.

Halfway across to Oakland a last sight of her gleaming sides was caught between two islands. The ensign and both signals dipped again, three times; her owner and his guests stood with uncovered heads as the little brass cannon spoke a last good-by; and then a certain mistiness, not wholly of the sea, enshrouded her, the white sides grew bluish in morning haze, the big ferryboat ploughed on, and the Japan cruise of the Coronet was a thing of the past.

> "Love, good-night, must thou go?
> When the day and the night need thee so!
> All is well; speedeth all to his rest."

"Taps."
[Extinguish lights.]

CHAPTER XXXIII

BACK TO AN ARIZONA COPPER MINE

Such is the patriot's boast, where'er we roam,
His first, best country ever is at home.
          GOLDSMITH.

"BISBEE washed away by a cloud-burst."

This cheerful headline, in letters much taller than necessary, on the first page of the San Francisco papers, had greeted our arrival from the Orient. Such was the reward of our search through the daily press to discover who were the actual nominees for President, and how the country at large regarded its two choices. Since at Bisbee, Arizona, is located a famous copper-mine with costly machinery, in which our host, the Coronet's captain, had far more than a general interest, and since the town is merely an outcome of the mine, its houses owned by the company, its inhabitants the underground workers with their superintendent, doctor, clergyman, and librarian, the washing away of his whole community was not only rather startling, but very moving to the Captain's emotions.

For a day or two telegrams had failed to get through, but at last the welcome message arrived from the superintendent, "Bisbee safe — no one killed." And when, as soon as possible after necessary business was finished in San Francisco, we found ourselves ensconced in the Buenaventura, and rolling over the arid deserts of southern California, the probable condition of the "works" was in the minds of the company almost as much as in that of the Captain.

In the Yuma desert a bush is not an incident merely, but an epoch. Miles of sand and reddish soil stretch away to barren mountains whose rough outline and scarred sides were made beautiful and ethereal by exquisite shadows and lights under a pale-blue quivering atmosphere. The thermometer stood at 100° F. in the car, — a rather lower point than usual in this region; but the air was so dry that it was by no means unbearable. Dust, however, sifted in through double windows, and powdered the little parlor.

A tempting mirage often appeared, — tantalizingly perfect presentments of ponds reflecting the hills, even hummocks of grass and rough rocks along the shore. Several times we should have been sure actual water lay at hand, except that it rose and flooded the country around some little station perhaps just passed, whose actual pitiful dryness had been, five minutes before, a sad contrast to the falsely rippled surface of that surrounding lake, now lapping gently its wood platform. Perhaps it was some similar appearance which caused the godfathers of the region to name the stations with cruel inappropriateness Sweet Water, Bubbling Spring, Running Brook, and the like, — pathetic sarcasm. So far as concerns fertility or moisture the whole scene might well have been a landscape in the moon.

But I trust there are no tramps in that desolately celestial region. The whole country in this earthly counterpart became more and more infested with that undesirable parasite. A gypsy camp, passed toward nightfall with flickering fire and picturesque figures about it, was decorative; but tramps cannot, by any possible stretch of imagination, be ranked in that category. They clung to the trucks beneath, stealing rides of a few miles at imminent risk to life and limb; they climbed to the roof of the Buenaventura, and were continually dislodged, even from our own observation platform. At every station beggars whined for food or money, the climax being reached at Yuma, on the banks of the turbid Colorado, where Indians, Mexicans, strange dialects, and mosquitoes swarmed in the hot evening air.

Leaving the Southern Pacific the following day, the Buenaventura was attached to a special engine for fifty or sixty miles' run (on the Arizona and South Eastern, a railroad belonging to the mining company) across a curious country, to Bisbee in the extreme southeast corner of the territory. The landscape, only less bare than the desert, showed yucca and century plants, varieties of flowering bushes here and there, and brilliant blossoms among the sparse grass, — the mesa covered with cattle, and bounded by superb purple mountains on every side.

Riding on the engine was attractive, but upon the cow-catcher even more so, — a species of luxury seldom allowed on through lines. Here, comfortably established on cushions, our feet resting upon the timbers joining in a point below for convincing argument with obstacles upon the track, the whole wide scene was most advantageously viewed.

But the track seemed to be, of all the windy mesa, the favorite reclining ground of herds, and too much slowing down out of regard for our safety became necessary, as the cattle calmly persisted in remaining until the engine was actually upon them; so after a few miles we reluctantly abandoned our seat on the pilot. Then with all lawful notice in shrill whistling, cows had to take their chances. Bleaching skeletons beside the track attested an occasional insensibility to warning; and a ruined town raised despairing chimneys to the silent sky, its adobe walls roofless and crumbling, still known as Charleston.

AINU WOMAN WEAVING ELM-FIBRE INTO CLOTH

Barren and more barren grew the country, — the soil more brilliantly red; then the track entered a narrow cañon, constantly more contracted as high mountains crept closer together. Whiffs of smoke appeared, tiny adobe houses straggled up steep red hillsides, themselves scarcely different in color; then tall chimneys and pipes spouting greenish vapor became frequent, and the car stopped. Tiers of houses clung to cañon walls, winding pathways connecting them wherever foothold could be seized, each as rough as the washed bed of a rapid torrent; across an innocent looking stream at bottom of the gorge a bridge leading to the intact and uninjured works; an enormous smoke-conductor eight feet in diameter and seven hundred feet long, lying at a steep angle up the mountain and pouring its incessant volume of sulphur smoke off, far above the little town, — this was Bisbee. No growing thing, not even the hardy cottonwoods, can live in the sulphur-laden air, even with the worst of it now carried off by the great flue.

This increase in sulphur, though greatly relieved, brought death to the few shrubs of Bisbee, and the air cannot, even now, especially in certain directions of the wind, be called favorable to agriculture.

As the Buenaventura lay comfortably sidetracked in the unique village, Mexican women, picturesque in black *rebosa*, their beautiful dark eyes looking at us curiously from swarthy faces, flitted by, and uncounted nationalities among the miners passed and repassed. We found later that twenty-two nations were represented (nearly everything but Mongolian), among them English, Spanish, Indian, South American, Welsh, Cornish, Irish, African, Norwegian, Swedish, Russian, Italian, Polish, Portuguese — and that fifteen languages were spoken.

Over seven hundred men are on the pay-roll, making, with their families, a population of about three thousand, all of whom are personally known to the management, which exercises so much kindly authority that Bisbee is an ideal mining community. Lawlessness is checked at once. A large store supplies at reasonable prices all needs, from white silk parasols and sets of silver-plated ware — both greatly in demand — to Navajo blankets, Mexican saddle-bags, and steeple-crowned hats. Two physicians, employed by the company, look after the general health; a devoted Welsh clergyman nourishes their souls; a fine library and reading-room are skillfully administered by a graduate of Pratt Institute; and the manager with his family are the good angels of the region.

It is a unique spot, the works with their pipes and puffing steam and smoke, coke heaps and slag piles the most pervasive element of the narrow cañon, while hundreds of feet above, on all sides, rise steep mountains, red, barren, mighty.

The trickling stream can without a moment's warning become a devastating torrent. Just before the Coronet landed, that cloud-burst in the mountains above, of which the papers told, had raced down the valley and Bisbee was nearly engulfed. Water rushed into the Copper Queen mine, over the floor of the works, and only stopped short of serious damage. No lives were lost, as the manager had telegraphed, but in a town farther on six or seven persons were killed. Hailstones broke half the Bisbee windows, even denting and riddling iron roofs.

Of the sudden, overwhelming power of cloud-bursts we were destined to see a thrilling example before the few days' visit was over.

At first sight one would say, "Machinery, arid precipices, sulphur fumes, no vegetation — terrible!" But a fascination not to be explained grows upon the stranger, partly understood as day after day passes in the little town, — fourth in importance and size in Arizona, a territory as large as New York and New England together.

The manager's pretty house stands at the top of three or four stone terraces, upon which by constant care a thick mat of Bermuda grass was green; oleanders were still living, and a vine or two over the piazza, while century plants and yucca, quite in their native habitat, gave an attractive air to the pleasant home where so much kindly hospitality kept open house.

Everybody was anxious to see the great mine; and in the afternoon the men of our "party of notables," as the Bisbee paper put it, charmingly arrayed in blue overalls from the store, the ladies in brown linen, boarded the elevator, and dropped into four hundred and twenty feet of darkness. Passing each level, an electric light gleamed for a moment. Once at the bottom, each guest with a candle investigated in Indian file the long corridors cut in the rock, through which little tracks are laid for cars to take out the ore. On both sides were rivulets of water from the late flood, and the procession proceeded cautiously, tiny candles flickering hither and thither in the turnings, as we followed our guides, the glimmer of whose lights, far ahead, showed the way. Stepping aside for filled cars to pass, while slowly trickling drops from above tinkled into pools below, soft white fungus clinging here and there to the walls, we kept on, seeing occasionally a rare bit of lovely light blue crystal, from drippings rich in ore.

When the Copper Queen was first opened the ore was very beautiful, abounding in crystals of sapphire blue called azurite, in delicate green, and malachite. Very rich, too, it proved, containing between twenty-five and sixty per cent. pure copper. But as mining went on, this particular variety grew more scarce, as well as bits of native copper.

But with the more prosaic material now mined the Copper Queen is still very rich, and not less than thirty tons of pure copper are shipped daily. In the ore are found not only sulphur, but traces of gold and silver, silica, lead, iron, zinc, antimony, arsenic, and other materials, all of which are blown off, or sent off, or turned into slag, except the bits of gold and silver. Never less than eight tons of pure copper average from one hundred tons of rough ore, while the early blue averaged forty, — another rich variety being jet black.

Walking across a plank over an apparently bottomless pit, and reaching a chamber too low to stand upright, candles held close revealed a fairy grotto. The roof and sides were of softest green moss like velvet, so delicate that a finger-touch brushed it away — and every leaflet of rich copper. Another cave, but vast and mysterious, was explored. Lofty and full of superb stalactites like alabaster, small apartments at the sides glitteringly splendid in the moving lights — this magnificent cavern, calm in the undisturbed repose of centuries, lay in the mountain's heart unknown, until a sudden blast accidentally opened an entrance to its gloomy wealth.

Impressive as were all the underground rooms and passages, and the ceaseless energy of labor above and below, the works at night were incomparably more so.

After "roasting" in a sort of rotary machine, the rough ore is dumped into four great furnaces together with a lot of coke (in the largest furnace two hundred and thirty tons of ore go in with fourteen of coke), where it is burned until "done," becoming liquid enough to run off. It is then two materials, — *matte*, containing copper, and useless slag. The latter, being lighter, rises, and is led out of the furnace at a higher level than the matte, which pours out its red-hot stream below. At this stage the matte is about fifty per cent. copper, thirty-five per cent. sulphur, and fifteen per cent. iron.

The matte left to cool is later put through a second furnace, from which it pours in streams of red-hot liquid flame into the two great Bessemer furnaces. In other words it is "Bessemerized" for about forty-five minutes, air being forced through it by a pressure of sixteen to twenty pounds to the inch. The sound is like a hundred engines together, and the flame, as it shoots up and out into the hood for carrying off fumes, is all shades of blue and violet and shining yellow, the swarthy figures in attendance knowing instantly by the color when all sulphur has gone. The enormous, seething caldron is finally tipped over, now a white-hot, indescribably glowing mass, the "cream" (slag) runs off the top, — a stream once more of liquid fire, pours itself into great vats on wheels, and is rolled away. When all the slag, chiefly iron, is poured off, oblong pots on wheels come in on the same track, receive the copper stream, now ninety-nine and three tenths per cent. pure, tumbling down in a cascade of glory, and roll off, — each bar, when cooled, weighing three hundred pounds, each heat usually yielding thirteen bars.

Men at work in the glare stick iron spikes continually through holes in the back of the converter, that passages for air-blasts may not become clogged, and when the red-hot or white-hot streams light up their faces, while showers of sparks fly off in wide-spreading masses, the effect is superbly weird.

The great Bessemers are lined with a sort of clay, which is constantly watched, lest it burn too thin, — too near the iron. When this happens it is wheeled away for the lining to be burned out. Six are always in use — two full of the copper, and four being burned out with radiantly lovely colors.

But something more, no less magnificent, was

yet to be seen. The slag, in its great iron pots on wheels, is run upon a small open train outside, men standing about amid the pots of red-hot slag, as spectacular as a scene in a theatre, and an engine, the "Little Queen," hastens off with it upon a tiny track to the slag-heap, a quarter-mile away. The molten material may have cooled a bit on the surface during its journey, flecks of dark crust dotting the red, but as each pot is dumped over the edge, to the valley one hundred feet below, it strikes the brink of the precipice and breaks or flows apart into a thousand semi-liquid fragments, which unite again in a glowing mass of incandescence, a rushing cascade of fire. The whole scene about these works at night is quite beyond adequate description.

Horseback rides by day over the barren mountains are as distinctive in their way. The animals are so trained to peculiarities of the region that they dash along at full canter up the dry beds of streams, along trails where a man could scarcely find footing, or straight up open hillsides to gain a short cut, leaving the washed-out roads to their own devices.

Bisbee itself is five thousand three hundred feet above the sea, and Juniper Flat, where a memorable horseback ride was taken, leads one up and ever up, seven thousand five hundred feet in elevation. Away from the works and their sulphur the air is extraordinarily clear and invigorating, the views extremely grand, over piled-up masses of red mountain peaks, with chasing lights and shadows and ineffable blue haze. At the divide the road descends toward Tombstone, and the view down the cañon was peculiarly beautiful, — even without vegetation of any sort except an occasional cactus, or the "mahogany of Arizona" (manzanita), an infrequent juniper or cypress, and ubiquitous yucca.

A solitary Indian on a mule, above us, was picking his way still upward. We seemed already perched upon the very backbone of the world, but a still wider range opened a few feet beyond, — far into the sunny Sulphur Springs Valley, with a bit of the blue San José Mountains of Mexico peeping over, and the Cananeas in the distance. Lookout Mountain, where scouts or sentinels used to watch for Indians during the Apache troubles, was sharp and distinct; Dixie Cañon and a dozen imposing peaks filled the horizon, — a tumble of mountains not unlike that seen from Pike's Peak.

A file of mules laden with firewood from some distant cañon passed us, driven by Mexicans.

Coming back to Bisbee, an exciting race with a mountain thunderstorm took place between elements and riders, — the black cloud and rushing drops barely behind all the way; but thanks to our sure footed-horses in their wild homeward gallop, the car was reached just as the first drops fell.

Down the stream from Bisbee, out on the free, breezy mesa, cantering without regard to road or boundary possesses a new charm all its own. There, breathing for the first time seems legitimately accomplished. Indefinite miles in extent, it is inclosed only by blue Mexican mountains of San José and the Sierra Madres on the horizon, the nearer Mule Mountains, and toward the north the Huachuca, where an army post is stationed among cañons of much luxuriance. In the vast plain are two small hills, one called Deer Point, where not long before a stage was held up by cowboys, and two men killed; farther, the Lookout Mountain, already once seen, with its strange castellated top. Cattle roamed at will over the great plain, now and then succumbing to thirst, as occasional bleaching skeletons and skulls suggested; coyotes ran ahead of us, jack rabbits and "cotton tail," and flocks of quail, among the scanty vegetation.

Yucca, and the mescal, from which a sort of whiskey is made, were the chief plants, but mesquite and bits of fluffy clematis, and more or less *ocatillo* were seen,—a curiously branching shrub covered with closely growing leaves. Settlers cut stakes from it for fences, but in the spring it suddenly sprouts, and lo! the most prosaic is possessed of a beautiful hedge full of scarlet blossoms. A white pillar marks the Mexican boundary, sole suggestion of proprietorship in the whole wide scene.

Another race with a storm, majestically sweeping up and completely hiding the Huachuca Mountains in its blue-black shadow, brought us back at twilight just in time to escape a fierce pelting with hailstones, and to see the fiery cascade of slag from below, leaping down the precipice in whirling sparks and flames, like molten lava, in a redly widening stream.

Once a storm had the advantage, — Arizona cloud-bursts were amply illustrated. A short ride down the stream, and a dark cloud seemed suddenly spread quite over the cañon; a few drops of rain fell, and hastily fording the shallow brook we rode the horses for shelter into a rough shed on the other side. In less than two minutes a wild downpour had shut out the sight of everything in a wall of descending water, and the innocent brook was a mad swirl of turbid, angry waves, — a foot, two feet, three feet deep, widening as we watched, deepening with every breath beyond a possibility of recrossing. The shed was slight shelter, open on three sides; hail and rain drove completely through it, while a small ravine between shed and house turned into another rushing stream which in a moment could not have been crossed. Fastening the horses hurriedly, it was the last possible opportunity to jump over the second stream on two or three stones still left uncovered. Scarcely had we gained the house when the last stone disappeared, and the frail dwelling on a tiny point of solid land was almost surrounded by yellow-red, deafening, foaming torrents, constantly more furious, and closer to the little porch with each moment. Rain still came down in sheets — above and on every side a watery wilderness — with a deafening roar.

In an hour the sky cleared. In another, the smaller stream had shrunk sufficiently to expose one or two stones, on which with the aid of a board from the good people who sheltered us we crossed, proceeding carefully on foot along the steep bank of the principal stream, still not less than twenty feet wide, finally reaching the railroad bridge at the village and the car. White hailstones lay about in heaps, and the cañon was an imposing sight.

Washouts detained the Buenaventura for a day or two, which started at last with considerable caution and slight speed. The whole Southern Pacific road was so washed and flooded that great lakes lay along the track, and the car rolled about as if we were once more at sea. The entire country was soft and muddy and spongy. Poor and squalid adobe villages on one side — on the other for an instant a distant view of the southern end of the Rocky Mountains, snow-covered and gleaming, and onward from El Paso we rushed, as if in very truth the train did

> ... "lap the miles
> And lick the valleys up."

Quaint old San Antonio; the lush forests in parts of Texas with birds still singing and armies of butterflies fluttering like brown leaves in autumnal gales; woods hung with solemn gray moss; the cotton fields and sugar plantations of Louisiana, its low-drooping trees and water-plants; New Orleans with its combination of modern cleanliness and beauty, ancient life and old French charm; the great Georgia cotton fields all in fluffy white; the distant Blue Ridge and changing foliage of Virginia; Washington welcomes, and more autumnal glories; farther welcomes in New York — with these the story of the Amherst Eclipse Expedition draws to its close.

But reunions of the participants have not been infrequent, and during the winter following, the freight steamer came through Suez with the apparatus; the Ainu collections were opened and displayed in scenes far different from those which witnessed their gathering; in February the Coronet reached her nest in Tebo's Basin, one hundred and fourteen days from San Francisco, completing her fourteen months' cruise of forty-five thousand miles.

The pink velvet has been restored to her guest-room walls, and the entire interior is refitted and furnished after her wanderings. One of the bottles, thrown overboard from the Coronet on the 27th of September, 1896, in latitude 43° N., longitude 135° 25' W., came ashore at Ross Bay Beach, Victoria, on the 1st of April, 1897; and his Imperial Majesty, the Emperor of Japan, in recognition of founding the Esashi library and services in the cause of education in northern Yezo, has conferred upon Professor Todd the imperial *sake* cup with its famous "go-shichi-no-kiri" crest in gold, and an accompanying document or diploma.

The heavens remain; sun and moon still pursue their steady cycle, and the astronomer patiently waits and works for still another eclipse. His life is a consecration to the best and highest. His joy over one new fact wrested from sun or star is more than the mere merchant's over an additional fortune made. He must possess the potentiality of a hero, the calm of a philosopher, even the uplift of martyrs of old. What wonder that he lives in startling nearness to the gigantic forces of nature and

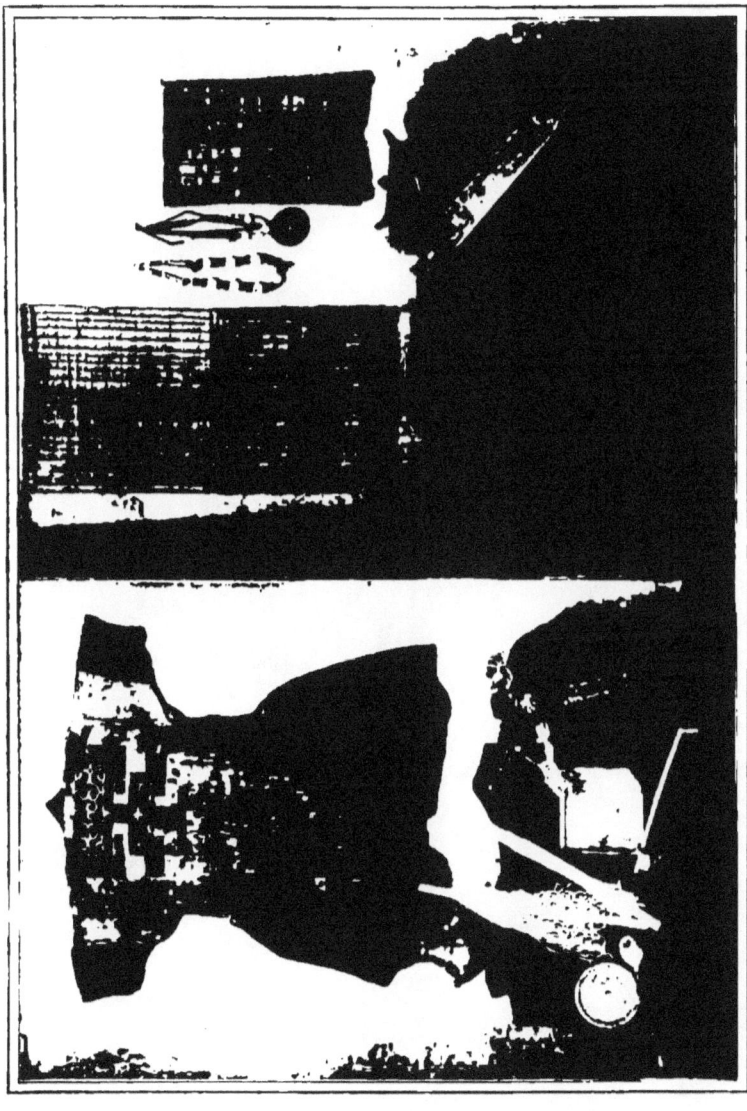

ARTICLES OF AINU MANUFACTURE.

Elm-fibre robe, god-sticks, head-dresses   Carrying-mat, aprons, bead necklaces, loom

their inconceivable operation? That in his clear eyes personalities, luxuries, and fashions, hates and envies, seem very small, and farther away than the stars he loves?

He often knows "the finer grace of unfulfilled designs;" but his hope springs perennial.

In cosmic spaces shadows cannot fail to fall, and the solid earth must now and then intercept them. Somewhere they will be caught, beneficently falling through unclouded skies.

INDEX

ADAMS, GOVERNOR, 94.
Adams, U. S. S., 56, 103, 123.
Adriance, Dr., v.
Ai (trout), 192.
Ainu (or Aino), 256, 259, 292; belief about eclipses, 321; collections, 259, 291; dread of death and ghosts, 315; fishermen, 282; hairy, 159, 244, 255, 293; house, first visited, 266, 267; legends, 264, 287; men, 268; salutation, 301; women, 268.
Aioina Kamui (Ainu Adam), 264.
Akadama (elm), 298, 299.
Aki province, 223.
Akkeshi, Yezo, 158, 160, 275, 324.
Albatross, 30, 33.
Aldebaran, 127.
Aleutian Islands, 351.
Alexander, Professor, vii, 108.
Alfred "the Great," 23, 132, 185.
Algaroba (Hawaiian tree), 46, 47, 109.
L'Alger, French cruiser, 149, 160, 235, 236, 277, 335, 336, 338, 345.
Aloha (Hawaiian farewell), 94, 124, 131.
Alpha Centauri, 126.
Amakura (heaven), 138.
Amherst, 106, 107, 205; cheer, 14, 22, 124; College Glee Club, 14; colors, 33, 237; preparation for expedition, 277, 278, 279.
Amur river region, xxi.
Anatomy of Wit, 327.
Ancient Mariner, 125.
Andrew, mate of Coronet, 129, 231, 232.
Antares, 135.
Aomori, 143, 171, 231.
"A 1" (private car), 15, 23, 32.
Apache troubles, 370.
Armstrong, General, 107.
Astrology (in Hawaii), 54.
Astronomer Royal of England, 160, 343.
Atlantic Monthly, The, viii.
Atsu (elm fibre), 298.
Awaji, Inland Sea, 217, 227.

Baden-Powell, Sir George, xix.
Baker, Mount, 20.
Baldwin Home, 117.
Ball, Sir Robert, 241.

Bandaisan eruption, 143.
Baseball in Yokohama, 180.
Bausch and Lomb, opticians, 278.
Bear killing, 309.
Bearskins, 260.
Beauty of Glazenwood, 26.
Benten, Japanese goddess, 218.
Beppu, Inland Sea, 226.
Bessemers (at copper mine), 368.
Betelgeux, 127.
Bisbee, 357, 360 ff.
Bishop, Bernice Pauahi, Museum, 53; Hall of Science, 107; Hon. Charles R., 53, 105, 107, 108; Mrs. Bernice Pauahi, 53.
Black Current (Kurosiwa), xxxiv, 137.
Blonde Frigate, 64.
Blow-holes, 75, 79.
Bluff, the, at Yokohama, 140.
Board of Health (Hawaiian), 112, 113, 114, 116, 120.
Boki (Hawaiian chief), 104.
Bonita (pilot boat), 29.
Boutet, Captain, 235, 277, 280, 334, 336, 338, 339, 343, 345.
Braemer, S. S., 5.
Brashear, Mr., optician, 276.
Buddha, Kamakura and Nara, 209.
Buenaventura, private car, 358, 359, 361, 373.
Bund, the, at Yokohama, 140, 149.
Burckhalter, Mr., astronomer, 160.
Burke, 254.
Burton, Professor, viii.
Byron, Lord, 64.

Cananeas mountains, 369.
Cape Horn (of Japan), 234.
Captain's birthday, 133.
Cascade mountains, 20; tunnel, 18.
Castle, Hon. W. R., 109; Mrs. S. N., 109.
Caves, burials in, 64.
Celestial Love, The, 229.
Century Magazine, The, viii.
Chabot Observatory, 160.
Chambers, 241.
Cha-no-you (tea ceremony), 151, 152.
Characters, Chinese, 242.
Chess, 31, 131, 232.
Chicago, 15.

Chief's journal, 229.
China, treaty with, 225; war with, 154.
Chinese, the, 62; compradores, 153; war, mementos of, 258.
Chipkommoi (sun), 309.
Christie (W. H. M., Astron. Royal), 160, 345.
Church (Central Union at Honolulu), 103; (Native, at Honolulu), 56.
Cingalese at Grand Hotel, 179.
Cleghorn, Mrs., 64.
Clerke, Miss, 242.
Clock, driving, 11; glycerine, 12; sand, 12.
Cloisonné, 178, 201, 202.
Coleridge, 125.
College, Agricultural, at Sapporo, 272.
"Colors," 150, 349.
Commutator, electric, 11.
Cook, Captain, 63, 64.
Copper, process of purifying, 366, 367.
Copper Queen mine, 363, 365.
Cormorant fishing, 188, 189.
Cornwall, Barry, 343.
Corona, xiii, xiv, xviii–xx, 8–10, 160, 277, 285, 320, 323–25.
Coronet, 4, 8–10, 13, 24–31, 45, 47, 56–59, 103, 134, 182; built when, xx; library, 36; log, 6; melody, 150, 349; saloon, 3; signal letters, 40.
Costume in Japan, 160.
Courlon, Captain Le Bouleur de, 277.
Crehore, Mrs., viii.
Crosby, Captain, 2, 24, 355.
Cross, Southern, 47, 123, 126, 133.
Cryptomeria (Japanese cedar), 186, 194.

Daikichi, 225.
Daikon (radish), 165, 173, 230.
Damien, Father, 119.
Dan-no-Ura, 224.
Dashi (float or car), 207.
Dauntless, the, xxiv, 3.
Deer Point, 370.
Deslandres, Professor, 160, 277, 280, 330, 332, 337, 343, 345.
Detroit, U. S. S., xxxii, 149.
Diamond Head, Honolulu, 40, 47, 124.
Dickinson, Emily, 58.
Dixie Cañon, 370.
Dodge, Mr., 59, 60.
Dole, Rev. D., 105; Mrs., 49, 50; President, vii, 45, 105, 109, 123.
Dole Hall, 105.
Doshisha (College), 204, 241.
Dryden, 68, 97.
Dumas, Midshipman, 277.
Dun, His Exc. Edwin, American minister, 169.
Dunkards, 22.
Dutton, "Brother," 118.
Dutton, Captain, 61.

Earthquake wave, 251.
Eclipse, apparatus, 5; beginning, 320; phenomena, 322; selecting station, xx, 158; tracks, xiv.
Elepaio (Hawaiian bird), 86.
Elm fibre as thread, 298.
El Monte, 25.
El Paso, 323.
Emerson, 125.
Emerson, Dr., 114, 117.
Emperor of Japan, 3, 144; message, 331; portrait, 238–39, 284, 331; unveiling of, 331.
Empress of Japan, 144.
Era of Meiji, 140.
Eri (neckerchiefs), 213.
Eruptions (in Hawaii), 74; (1868), 72, 75; (1880–81), 73, 75; (1892), 74.
Esashi, 158, 160, 171, 217, 234, 241, 243, 252, 256, 271, ff. 375.
Etchuya Inn, 258, 261, 262.
Expedition, 35, 36, 170, 171, 229, 234, 236–40.
Expeditions of different nations, 160.

Farallones, 29, 354.
Field, Kate, 63, 97, 98, 100, 101, 103.
Fiji Islands, 149.
Fisheries in Yezo, 233.
Flag-ship U. S. S. Olympia, 149.
Flathead River, 17.
Floats, Kyoto, 207, 283.
Floriponda, 82, 84.
Flying-fish, 34.
Folk lore story, 93.
Formosa, Governor of, 161.
Fort Peck Indian reservation, 15.
Fortune-telling in Yezo, 314.
Fourth of July at Yokohama, 179.
Francis, Mr., v, vi.
Friedländer, Dr., 59.
Fuji, 137, 138, 140, 161, 182–84, 186, 268, 349.
Fujino tea-house, 225.
Fujita (Hundred Steps), 148.

Gaisen (dance), 167.
Gardens, Imperial, 161.
Gay, 264.
Geisha (Gifu), 190; melody, 191.
Gerrish, Mr., v.
Geta (shoes), 155.
Gifu, 188.
Go-downs, 246.
Goerz, optician, 278.
Gohei (paper prayer), 218, 244.
Golden Gate (San Francisco), xxxvii, 28, 349, 355; (Uyeno), 174.
Golden Pavilion, 198.
Goonies, 30, 32, 33, 39.
Government, Japanese, 156, 159.
Great Northern Railway, vii, 14, 16.
Guest book, 344; at Fujita, 147.
Gundlach Optical Co., 278.

INDEX 379

Hachinoye, 143.
Hakodate, 159, 170, 171, 237, 243, 253, 254-57.
Hara, His Exc'y, Governor of Hokkaido, vii, 233, 275.
Harte, Bret, 155.
Hawaii, annexation, 45; bride, 92; climate, 110; flag, 122, journey, 68; language, 44; leaving, 95; lepers, 111; melodies, 44, 47, 54; minister from, 46; politics, 45; relics, 53; roadsides, 83; sea coast, 65; singing, 100; spirit of modern, 87, 109; volcanoes, 58; women, 54, 56.
Hawaiians, 42, 122, 199.
Hawthorne, 211, 287.
Hayasbi, Mr., vii.
Heijo, 210.
Helmets of feathers (Hawaii), 91.
Hemans, Mrs., 241.
H. Henry VI., 188.
Henry Gandell's Leap, 75.
Herbert, 181.
Herod, Mr., vii, 170.
Hibachi (brazier), 188, 206.
Hill, President, Great Northern, vii, 14.
Hills, Captain, 160, 343.
Hilo, Hawaii, 73, 82-84.
Himiongami, 213.
Hiroya, Mr., 290, 335.
Hohei-kwan, Sapporo, 258, 261.
Hokkaido, 273, 274, 294; governor of, 233, 275; observations in, 157; oyster beds and fisheries in, 233; wealth of, 282.
Holoku (Hawaiian dress), 42, 61.
Hongo, 250.
Honolulu, 3, 12, 40-43, 46, 57, 66, 90, 104, 110, 129.
Horn, Cape, 2, 4, 24, 106, 129; of Japan, 234.
Horseback riding, 49, 285.
Horses in Yezo, 286.
Hosmer, President, Oahu Coll., 107-9.
Huachuca mountains, 370, 371.
Hualalai, 95.
Hula (dancing girls), 75.
Hundred Steps, tea-house, 147.
Hurbin, Captain, 277.

Ideographs, 145.
Ieie (vine in Hawaii), 82.
Inao (god-sticks), 297, 302, 313.
Independent, The, viii.
Indians, 16.
Inland Sea, xxxii, xxxiii, 159, 180, 217, 229, 241.
Instruments, 8, 9, 132, 278.
Ioi (Hawaiian flower), 89.
Ishikawa-maru, 284.
Ito, Count, 225.
Iwalani, S. S., 119, 121.
Iwate, prefecture, 143; branch of Red Cross, 144.

Jacula Prudentum, 181.
James, A. C., iii. v, xxiii, 21, 346; D. W., vi, 3; Mrs. A. C., v, 36.
Janssen, xviii.
Japanese, 62; alphabet, 211; dinner, 165; landscape gardeners, 84, 161; national anthem, 150, 154, 239, 330.
Jiji, 144.
Jinrikisha rides, 140, 144.
Jonson, B., 327.
Juniper Flat, 369.

K D J B (Coronet signal letters), 40.
Kaahumanu, 55.
Kaawaloa, 63, 97.
Kabayama, Mr., vii, 205.
Kabuka, 269, 270.
Kago (palanquin), 225, 285.
Kagura (heavenly music), 213.
Kahuku, Hawaiian town, 72.
Kahuna (witch doctor), 76.
Kaiana, Hawaiian chief, 51.
Kailua, Hawaiian town, 51, 63, 99; relics, 94.
Kaiulani, Princess, 45.
Kalakaua, King, 3, 55, 74, 95.
Kalaupapa, old town on Molokai, 120.
Kalawao, 120.
Kalo (vegetable), 51.
Kamaishi, town in Japan, 246, 250.
Kamakura, Daibutsu at, 210.
Kamehameha I., 51, 53, 74, 93, 94, 99.
Kamehameha II., 100; IV., 55.
Kamogawa, river in Kyoto, 197.
Kanaka, island native, 62.
Kanda, Professor, 205.
Kant, Immanuel, 128.
Kapapala, Ranch, Hawaii, 71, 89.
Kapiolani, 70; Queen Dowager, 95.
Kasuga, temple at Nara, 211, 212; god of, 215; dance at, 212.
Kealakeakua Bay, 63.
Keats, 42.
Keauhou, Hawaiian town, 100.
Keopuolani, Queen, 55.
Kewalo reefs, 106.
Ki (leaves), 90.
Kidda (Ainu, for star), 309.
Kilauea, 59, 68, 70, 76, 78, 79, 89, 147.
Kilauea-iki, 81.
Kimono (Japanese dress), 142, 148.
Kimotsuki, Captain, 251, 273, 290.
Kinkasan, Japanese island, 143.
Kipling, 126.
Kitami, province of Yezo, 157, 158, 160, 170, 217, 255, 285, 335.
Kiushiu Island, 226.
Koa, Hawaiian wood, 108.
Kobe, 159, 180, 182, 183, 184, 221, 226, 228.
Kobo Daishi, 211, 213, 242.
Kochibe, Professor, 250.
Königsberg, xviii.
Kojiki, oldest Japanese book, 294.

Kokuzo Bosatsu, god of the universe, 210.
Kompira, temple, 218.
Komyo, Empress, 213.
Kootenai River, 17.
Korea, 149, 224.
Kori (basket), 185, 196, 262.
Koro-pok-guru, 293, 294.
Kotohera, Inland Sea, 218.
Krombi (water insect), 311.
Kuakini, 94.
Kuny Chipkommoi (the moon), 309.
Kupapa-u (a corpse), 26.
Kure, Inland Sea, 219.
Kurosiwa, xxxiv, 137.
Kuruma, or jinrikisha, 185.
Kurumaya (runner), 148, 178, 189.
Kushiro, Yezo province, 157.
Kwanko-maru, S. S., 263, 268, 270.
Kwannon, goddess, 199, 219.
Kyogen (a play), 167.
Kyoto, 183, 194, 197, 200.

Lahaina, Hawaiian town, 62.
Laiakanoe hale, Point of Mists, 48.
Lanai, island, 62, 90, 93; veranda, 46, 90, 95.
Langley, Professor, xix.
Lantana, 50.
La Perouse Strait, 269, 271.
Launfal, Sir, 111.
Lava flows, 70, 72, 73, 74.
Legation, United States, in Tokyo, 156, 170.
Lei (wreath), 44, 46, 47, 62, 75, 90, 91, 123.
Lepers, 111; band, 115, 116; expense of, 113; female quarters, 119; horses for, 121; marriages of, 112; seclusion of, 112; suffering among, 120.
Leprosy, anæsthetic, 117; evidences of, 115; first in islands, 111; germ found, 122; painless, 121; how transmitted, 122; white, of Syria, 122.
Lick Observatory, 160; party, 275.
Li Hung Chang, 225.
Likelike, Princess, 64, 74.
Liko lehua (Hawaiian plant), 89.
Liliha, wife of Boki, 104.
Liliuokalani, 55, 64, 74.
Longfellow, 216.
Lookout Mountain, 269, 370.
Loti, Pierre, 336.
Lotus, 174, 186, 199, 200, 209.
Lowell, J. R., 111, 194.
Lowell, Percival, 155.
Lukula, a Hawaiian prophet, 76.
Lunalilo, former Hawaiian king, 112.
Lyly, 327.
Lyman, B. S., 272.

MacArthur, S. S., 26.
McGrew, Dr., 103.
McNair, Admiral, 151, 179.

Maida, bombardment of, 224.
Makai (going toward mountains), 72.
Makudsu Kozan, famous potter, 348.
Manoa Valley, 48, 106, 109.
Manono, wife of Hawaiian chief, 54.
Maple Club, 161, 168.
Marlinspike birds, 39.
Mashika, 266.
Matsumæ, 328.
Matsuri, 205, 283.
Matsuyama, 226, 227.
Matte (copper), 366, 367.
Maui, island, 62, 102.
Mauka, leaving the heights, 72.
Mauna Kea, 66, 89.
Mauna Loa, 58, 59, 60, 64, 65, 68, 71, 74, 75, 76, 86, 89.
Meiji, present era in Japan, 140, 204, 252.
Merrill, Mr. (San Francisco), vii.
Miaki Island, 137.
Mikura Island, 137.
Mila Head, xxx.
Millochau, M., 160.
Mills, Consul, at Honolulu, 120.
Mills, President, 107.
Mississippi Bay, 140, 344.
Mitsui family, 144.
Mittau, MM., 160.
Miyabe, Professor, 258.
Miyagi, 143, 248.
Miyajima, 222, 223, 224; festival, 223.
Miyako-maru, S. S., 216, 217, 218, 220, 225, 226.
Miyanoshita, 343.
Moats, in Tokyo, 176; in Kyoto, 200.
Mokuaweoweo (crater), 58, 68, 74.
Molokai, Hawaiian island, 40, 62, 111, 112, 113, 114.
Mombetsu, Japanese town, 238.
Monocacy, U. S. S., vi.
Montgomery, 241.
Morse, Professor, 259, 291, 312.
Mother Carey chickens, 39.
Mule Mountains, 370.
Murakami, Mr., vii, 241, 243, 245, 252, 261, 263.
Museum at Sapporo, 258.
Mutsu, Count, 225.
Mykasa-yama, Nara, 213.
Mynah bird, 85.
Myorin Kwannon, 210.
Myoshinji Temple, 198.

Nagara, river at Gifu, 189.
Nagasaki, 339, 345.
Nagata, Professor, 316.
Nagoya earthquake, 143.
Nakadori, 177, 178.
Nakamura, Professor, 157.
Nakashima, Inland Sea, 222.
Namikawa (Tokyo), 178, 179; (Kyoto), 201, 202, 203.
Nara, 209, 210, 243.

INDEX

Naruto Channel, 228.
Nation, The, viii.
Neesima, Mr., 204.
Nemuro, province of Yezo, 157.
Newcomb, Professor, 242.
New England, 30, 70, 71, 83, 254, 287.
Nichi-Nichi, 144.
Nikko, 186, 343, 345.
Nishi Hongwanji, temple, 198.
Nishimura, 203, 204.
Nitobe, Professor, 233, 258, 259.
No dance, 167.
Nomamura, Inland Sea, 218.
Norway, xxi.
Nova Zembla, xvii, xxi.
Nozawa, Mr., 233, 235.
Nuuanu pali, 50, 51, 52.

Oahu, island, 40, 51; college, 104, 105, 107, 108; Glee Club, 108.
Oakland, 24, 356.
Obi (sash), 142.
Observations, meteorological, xxi, 156, 275.
Octopus, 33.
Odyssey, 30, 258.
Ogawa, Mr., viii, 170.
Oginohama, 244, 245, 251.
Ohayo ("good morning"), 281.
Ohia, tree with scarlet blossoms, 71, 75, 84.
Ohiyo, a kind of elm tree, 298.
Okachi, 248.
Okhotsk, Sea of, 159, 269, 274.
Okita, 184, 185, 196, 197, 201, 207, 221, 224, 345.
Olympia, U. S. S., xxxii, 149, 150, 179, 181, 182, 349.
Olympian Mountains, 20.
Omao (Hawaiian bird), 86.
Ondo, strait, 219, 220, 222.
Onivake, 269.
Onomichi, 219.
O-o (Hawaiian bird), 91.
Osaka, floods about, 243.
Osgood, 8.
Oshidomari, 269.
Oshima, Mr., viii, 233, 289, 290, 332.
Ota, "Wizard of," 348.
Otaru, 233, 243, 252, 255, 261, 266.
Outlook, The, viii.
Owl's Head, L. I., 2.
Oyama, Countess, 176.

Pacific, 9, 30, 31, 32, 40, 43, 136, 137.
Pahala, 69, 94.
Papaia (fruit), 56.
Paris Observatory, 160, 276.
Patagonia, 6.
Pauahi Hall, 105, 108, 109.
Peabody Museum, Salem, 259.
Pearl Harbor, 48.
Peeresses' School, 176, 177.

Pele, goddess of fire, 70, 74, 77; cave, 75; flower, 71, 75, 89; offerings to, 75; defiance of, in 1824, 70.
Pelican Island, 241.
Pemberton, J., v, vi, 229.
Pericles, 88.
Perry, Commodore, 147.
Phœnix Hall, 209.
Plate-holders, 10; endless chains of, 11, 36, 236.
Plays, old classic, 165, 166.
Plumeria, 46.
Poi (national Hawaiian dish), 51, 65, 66, 88, 91.
Poillon, Messrs., 2.
Point of Mists, 48.
Poison, in arrows, 311.
Polar axis, 11.
Polaris, 126, 133, 135.
Polynesia, 53.
Poronaibo, 295.
Portland, 22.
Portuguese, 43, 62, 85.
Puget Sound, 20.
Punahou, 104, 105, 107, 108.
Punaluu, 66, 73, 86, 88.
Punch Bowl (crater), 124.
Punkah, 170.

Queen, 92; Emma, 53, 55; Keopuolani, 55; Liliuokalani, 55, 64, 74.
Queen Regent, Kaahumanu, 55; Kinau, 55.

Rainier, Mount, 20.
Rebosa, 362.
Reibunshiri, 269.
Reporters, 21, 142, 355.
Reverie, 194.
Revolution, 176.
Richard barometer, 132.
Rishiri, 268, 269.
Roche's Point, Ireland, 2.
Roentgen rays in corona, 325.
Ronins, 175.
Rotation of corona, 160, 277.
Round Top, 196.
Ruth, Princess, 74.

Saghalien, 271, 274, 311, 329.
Sake, 165, 306; Sake Cup, 374.
Sakura (cherry blossom), 179.
Sakura-maru, 170, 171.
San Antonio, 373.
Sandy Hook, 5.
San Francisco, 2, 4, 7, 15, 21, 24, 25, 32, 69, 129.
San José Mountains, 369, 370.
San-ju-sangendo, 199.
Sankwan Island, 248.
San Rafael, 26.
Sapporo, 159, 170, 233, 255; Imperial Agricultural College at, 233, 255.
Saru-sawa-no-ike, pond in Nara, 214.

Sausalito, 7, 24, 25, 26, 28.
Sawayama, Mr., 205.
Schaeberle, Professor, 160.
Schoolhouse, Esashi, 280, 281; old, 275, 280; new, 228, 327, 330.
Scorpion, 126.
Sea of Japan, 269; of Okhotsk, 159, 269, 274.
Seattle, 20.
Seifu, famous potter, 201.
Sei-yo-ken, 173, 174.
Sendai, Bay of, 143.
Shakespeare, As You Like It, 104; II Henry VI., 7, 188, 327; King John, 318; Richard III., 209; Titus Andronicus, 24.
Shanties, 36, 37, 38, 134, 352.
Shasta, Mount, 22.
Shiba temples, 174.
Shijo, 214.
Shikoku, island in Inland Sea, 227, 242.
Shikotan, island of, 338.
Shimbun (newspaper), 142.
Shimidzu, 183, 184.
Shimonoseki, 224, 225.
Shinto festivals, 205.
Shirakawa, vi, 10, 243; apparatus at, 325; in 1887, 170.
Shirasaka-San, 331, 335.
Shiriya Light, 251.
Shizuoka, 185, 187.
Shodoshima, 218.
Shundoku (treasure box), 299.
Sierra Madras, 370.
Signals, communication by, 40, 182.
Skykomish River, 20.
Smith, Sydney, 139.
Smith and Terry, 2.
Some-San, 331.
South Sea Islands, specimens, 53.
Soya, Cape, 234, 255, 268, 271.
Spectroscopes, 8, 277.
Spica, 135.
Spokane, 17.
Sulphur caves, 81.
Sulphur Springs Valley, 369.
Suruga Gulf, 183.
Suruga-maru, 171.
Suzuki-San, Vice-Governor of Hokkaido, 371.

Tadotsu, 218.
Tairen-maru, S. S., 243, 244, 245, 251, 253, 255.
Takemikatsu Chi-nomikoto (god at Nara), 213.
Tamaiya inn, 188, 193.
Tamalpais, Mount, 26.
Tanabe, 147.
Tantalus, 124.
Taps, 356.
Taro, or Kalo, 51.
Taro-patch (stringed instrument), 47.

Tartary, Gulf of, 269.
Tattooing, 303, 304, 311.
Terao, Professor, 160, 276, 330, 332, 333, 346.
Thaxter, Celia, 1.
Thompson, E. A., v, 232, 278, 321; E. F., 24.
Ti (or ki) leaves, 90.
Tidal wave, 73, 143, 245-49.
Time bells, 149, 150.
Times-Herald, Chicago, 101.
Todd, Professor, v, vi, xxi, 10, 156-57, 233-37, 240, 278, 330, 337, 346, 374.
Tokaido, 185.
Tokonoma (recess), 187, 188.
Tokyo, 143; eclipse party, 276; Observatory, 160; Central Meteorological Observatory at, 157, 275, 319.
Tomo, 219.
Tonakai (deer), 329.
Toyoura, 224.
Treasures of the deep, 241.
Tsuda, Miss, viii.
Tsugaru Strait, 251, 252.
Tsukiji, Tokyo, 173.
Tubi, Inland Sea, 228.
Turner, Professor, 160, 343, 344.
Tuscarora Hollow, 250.

Uchimura, Mr., 205.
Uji, 209.
Ukulele, stringed instrument, 44, 47, 100.
University at Tokyo, 164, 273.
Urtica, fibre, 299.
Uyeno Park, 174.

Vancouver, 137.
Volage, H. M. S., 100.
Volcano House, 76, 77, 79, 81, 82, 89.
Vries Island, 137, 183.

Waa (canoe), 63.
Waianea mountains, 48.
Waikiki, near Honolulu, 46.
Waikolu, 120.
Wakkanai, 234, 271, 280.
War, with China, 145, 154.
Water-lemon, vine, 82.
Weather, Imperial Bureau of, 156, 157.
Wellington, Washington, 19.
West Africa, 10, 12, 326.
Wheeler, Mr. (San Francisco), vii.
Wilkes Scientific Expedition, 60.
Wright, Dr.; Yale Univ., 278.

Yaami, hotel in Kyoto, 194, 195, 196.
Yacht Club, New York, 2; San Francisco, 24; Yokohama, 149.
Yale University, 106.
Yamatoya, 348.
Yedo, 143; Bay, 139, 181.

Yezo, 11, 157, 171, 217, 251, 256, 272; bears, 244; evolved from chaos, 264; horses, 285, 286; shores, 252; traveling to, 241; west coast, 255, 265.
Yokohama, 3, 12, 144, 159.
Yonsike (an insect), 311.
Yoritomo, 300.
Yorktown, U. S. S., xxxii.
Yoshimitsu, 198.
Yoshitsune, 300.
Young, Professor, xix, 97.
Yuma desert, 358.
Yusen Kaisha, officials, 233, 266; steamers, 243, 255, 334.

www.ingramcontent.com/pod-product-compliance
Lightning Source LLC
Chambersburg PA
CBHW021424300426
44114CB00010B/633